F

Insurance
under the
JCT Forms

Second Edition

Frank Eaglestone

PhD, LLB, FCII, FCIArb, Barrister
Chartered Insurance Practitioner

Blackwell
Science

© Frank N. Eaglestone 1985, 1996
Blackwell Science Ltd
Editorial Offices:
Osney Mead, Oxford OX2 0EL
25 John Street, London WC1N 2BL
23 Ainslie Place, Edinburgh EH3 6AJ
238 Main Street, Cambridge
  Massachusetts 02142, USA
54 University Street, Carlton
  Victoria 3053, Australia

Other Editorial Offices:
Arnette Blackwell SA
  1, rue de Lille, 75007 Paris
  France

Blackwell Wissenschafts-Verlag GmbH
  Kurfürstendamm 57
  10707 Berlin, Germany

  Zehetnergasse 6
  A-1140 Wien, Austria

First published 1985 by Collins Professional
and Technical Books
Second edition published by Blackwell
Science 1996

Set in 10/12pt Palatino
by DP Photosetting, Aylesbury, Bucks
Printed and bound in Great Britain
by Hartnolls Ltd., Bodmin, Cornwall

DISTRIBUTORS

Marston Book Services Ltd
PO Box 87
Oxford OX2 0DT
(*Orders:* Tel: 01865 791155
         Fax: 01865 791927
         Telex: 837515)

USA
  Blackwell Science, Inc.
  238 Main Street
  Cambridge, MA 02142
  (*Orders:* Tel: 800 215-1000
           617 876-7000
           Fax: 617 492-5263)

Canada
  Copp Clark, Ltd
  2775 Matheson Blvd East
  Mississauga, Ontario
  Canada, L4W 4P7
  (*Orders:* Tel: 800 263-4374
           905 238-6074)

Australia
  Blackwell Science Pty Ltd
  54 University Street
  Carlton, Victoria 3053
  (*Orders:* Tel: 03 9347-0300
           Fax: 03 9349-3016)

A catalogue record for this title
is available from the British Library

ISBN 0–632–03969–8

Library of Congress
Cataloging-in-Publication Data

Eaglestone, F.N. (Frank Nelson)
    Insurance under the JCT forms/Frank
Eaglestone. – 2nd ed.
        p.   cm.
    Includes index.
    ISBN 0-632-03969-8 (alk. paper)
        1. Construction contracts–Great Britain.
    2. Construction industry–Insurance–Law
    and legislation–Great Britain.   I. Title.
    KD1641.E24   1996
    343.41'078624–dc20
    [344.10378624]   95-50735
                                         CIP

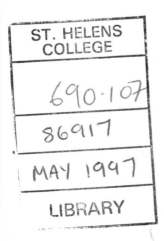

Contents

Preface to the first edition

The first edition of my book *The R.I.B.A. Contract and the Insurance Market* was published in 1964 and went to three editions. The last was published in 1969 (after the July 1968 Revision which had some effect on the insurance clauses), with a supplement in 1974. The last edition is now out of print, although there is still a demand for the book (the publishers are no longer operating in that field). Thus, with the advent of the JCT 1980 edition of the Standard Form of Building Contract, it would seem an appropriate time to write again on this subject.

The major aim of this book is to supply the insurance official, architect, quantity surveyor, employer, contractor and even possibly the lawyer with an analysis of the clauses which directly and indirectly concern the insurance industry. Furthermore, it is intended to give details of those insurance policies and bonds which are involved in providing the protection required (whether directly called for or not) by the clauses just mentioned. At the ends of the two chapters concerning the main insurance policies (Chapters 6 and 7) I have endeavoured to indicate where the basic insurance policy cover required by the contract (or considered advisable in the case of the works) does not protect the contractor and employer; each of these parties is considered separately.

It seems inevitable when an author embarks on a technical book involving contract terms that the representative bodies decide to introduce a new edition or revision or threaten to do so. It happened in 1974 when publishing the above-mentioned supplement, and the present time is no exception, for it is understood that some alteration to the insurance clauses is expected. For example, it is completely contrary to current practice for the JCT Contract to require a fire and special perils policy for the works when it is the usual practice to provide a contractors' all risks policy, bearing in mind the contractor's responsibility to complete the works. Hopefully this situation will be rectified.

I am very grateful to the following who assisted me in various ways; without such help the book might never have been completed: R.J. Bragg, J. Bridgman, C. Burgess, N.J. Burne, B. Edwards, D. Ford, E. Gamlen, R. Lewis, P. Madge and J. Tunnicliffe. Once again I have to thank Barbara Clegg for her work in typing from my manuscript.

Acknowledgements are made to P.H. Press Ltd and George Godwin Longman Group for allowing me to use certain parts of my other works. Also I appreciate the kindness of the management of the Allstate Insurance Company Ltd in allowing me to use their policies and other documents by way of illustration. Extracts from JCT 80, the Agreement for Minor Building Works, the Fixed Fee Form of Prime Cost Contract, the JCT Standard Form with Contractors' Design, NSC 4 and JCT Practice Notes are reproduced by kind permission of RIBA Publications Ltd, the copyright holder.

Frank Eaglestone
1985

Preface to the second edition

When the publishers approached me to bring this book up to date I had to consider whether there was enough new material, apart from the supplement, to justify this second edition. It did not take long to appreciate that, with growth in the last decade of case law and statutes in the legal liabilities field, and with new legal cases concerning the JCT contracts (within the area of this book) as well as those in the insurance world, a new edition was fully merited.

The first edition of this book came out in 1985, and in early 1987 the supplement concerning the amendment to the JCT 1980 edition was published. Since then there have been, among other matters, a line of cases bringing about the demise of the case of *Anns* v. *London Borough of Merton* (1977), the effects of the Latent Damage Act 1986, the Consumer Protection Act 1987, the advent of the Management Contract (JCT 1987 Edition) and the Measured Term Contract (1989 Edition), plus the revised procedure for the nomination of a sub-contractor by the JCT.

In the insurance world there has recently been the House of Lords' decision in *Pan Atlantic Insurance Co Ltd* v. *Pine Top Insurance Co Ltd* (1994) concerning the principle of utmost good faith and the meaning of a material fact. The JCT have produced their 1994 Guide to Terrorism Cover with its provisions in JCT forms on insurance of the works and existing structures and contents for physical loss or damage due to fire or explosion caused by terrorism. Finally insurers are placing a limit of indemnity of £10 million on their employers' liability policies in 1995, whereas in the past there has been no limit under such a policy.

Therefore in conclusion there is ample reason for a new commentary.

According to the RICS Contracts in Use Survey 1993, JCT standard forms continue to dominate the market. Whether taken by number or by contract value, JCT forms of one sort or another account for 80% of the sample.

When I wrote my first book, in 1964, about insurance under the RIBA contract, as it was then incorrectly called, the Banwell Report made the suggestion that there should be a common contract for all construction work. However, in 1967 arguments for and against a common form for building and civil engineering were published in the *Action on the Banwell Report*, and by 1974 the objective seemed to be more remote than ever. By this time the

JCT Standard Form, far from combining with the ICE Conditions, had sprouted additional contracts such as the Minor Works Form and since then there have been many variations of both the JCT and ICE contracts. Now we have the Latham Report published under the title *Constructing the Team* recommending a New Engineering Contract promulgated by the Institution of Civil Engineers which is suitable for use in the design and construction of new works involving any, or all, of the engineering and building disciplines. At least it has taken the step which was only an idea in the Banwell Report. However, whether the JCT will take up the Latham recommendation remains to be seen.

My thanks are due to John Goodwin of the Builders Accident for undertaking the daunting task of reading the whole script and making many useful suggestions. I also acknowledge the assistance I received from Noel Burne of Elrond Engineering, John Frost of the RHH Surety & Guarantee, Frank Leonard of Hamilton Leonard & Co, and Mel Walmsley, in reading various chapters and making helpful comments. I record the kindness of the management of the Independent Insurance Co Ltd in allowing me to reproduce their Contractors' Insurance Policy in Appendix 2, and the Chartered Institute of Loss Adjusters for extracts from *Contractors All Risks Insurance* reproduced in Chapter 7, parts of Chapters 9 and 10 and in Appendices 7, 8 and 9.

Frank Eaglestone
October 1995

Chapter 1

Introduction to legal liabilities

The wording of construction contracts inevitably widens the contractor's legal liability for damage to property (including the works) and for personal injuries caused by the carrying out of the works. Therefore it is necessary to set out the contractor's liability in tort and his more common statutory duties in order to see the extent to which the contract terms add to this tortious and statutory liability. In this chapter insurance will not be considered, but it is appropriate to point out that without insurance the Standard Form of Building Contract (JCT 80), in keeping with the majority of construction contracts, could not operate. While the main clauses considered in this book are clauses 20 to 22 inclusive, a wider field will be covered. In addition to discussing those clauses having some connection with the main insurance clauses, the other JCT contracts will also be reviewed in the light of insurance.

The following are the basic legal liabilities which exist in the UK and with which this book is concerned.

(1) Negligence
(2) Nuisance
(3) Absolute or strict liability
(4) Trespass
(5) Liability under contract
(6) Breach of statutory duty
    Although this is regarded by legal authors as a tort, and the liability here is 'strict' in that it can lie in some circumstances even though the conduct of the wrongdoer is neither intentional nor negligent, it is placed last in order to distinguish it from the common law for the purposes of this book.

The main torts } *Common law* } *Civil as distinct from criminal law*

The difference between a crime and a civil wrong depends on whether criminal or civil proceedings follow it. In the latter event it is a civil wrong. Sometimes the act is capable of being followed by both civil and criminal proceedings when it is both a civil wrong and a crime. The terms used are also different. In civil proceedings a plaintiff sues a defendant (pursuer and defender in Scotland) and success results in judgment for the plaintiff. In criminal proceedings a prosecutor prosecutes a defendant or an accused and, if the prosecution is successful, the result is a conviction.

Another difference between a crime and a civil wrong is in the burden of proof. The burden which is on the prosecution in the case of a crime is proof beyond reasonable doubt, whereas in the case of a civil wrong the burden on the plaintiff is to adduce sufficient evidence of fact to show on the balance of probability that the defendant was in the wrong. This difference was illustrated in the case of *T.D. Radcliffe & Co.* v. *National Farmers' Union Mutual Insurance Society* (1991) where the plaintiff's principal shareholder and director was acquitted of arson, but his insurers won their civil case when they refused to pay the resultant claim.

The phrase 'common law' has various meanings. In this book it means the law that is not the result of legislation, i.e. it is the law created by the custom of the people and the decisions of the courts. The latter is sometimes called case law, but there are case law decisions on breach of statutory duty, which has been defined earlier as exclusive of the common law. From this it will be realised that it is impossible to explain the common law without defining as one proceeds.

One of the main difficulties is to define 'tort', which is one type of civil wrong. From the preceding list it will be appreciated that a tort is a civil wrong independent of contract, i.e. it gives rise to an action for damages irrespective of any agreement giving the right to take the proceedings concerned. However, it is basic to the law dealt with in this book that, subject to the Unfair Contract Terms Act 1977, a term of the contract between the parties thereto may override a tort action otherwise available. The meaning of 'tort' will be appreciated when each of the main torts is considered in turn. In this connection it should be realised that there are defences to the following torts, of which the main ones are contributory negligence and consent to run the risk by the claimant, but the same defences are not always available to each tort.

The Law Commission's publication No 219 dated 9 December 1993 entitled *Contributory Negligence as a Defence in Contract* refers to the categorisation of contractual duties adopted by Hobhouse J in *Forsikringsaktieselskapet Vesta* v. *Butcher* (1986) and affirmed by the House of Lords in 1989. These categories are where the defendant's liability:

(1) arises from a contractual provision which does not depend on negligence on his part;
(2) arises from a contractual obligation of care which does not correspond to a tortious duty of care which would exist independently of contract;
(3) is the same in contract as his liability in the tort of negligence independent of the existence of any contract.

The Law Commission's commentary then concludes that the Law Reform (Contributory Negligence) Act 1945 only applies to actions in contract which come within category (3). Also the Commission considers that, although the law might have developed so as to allow apportionment in a wider category of cases, it is now clear on the authorities that such development is not possible under the 1945 Act. *Barclays Bank plc* v. *Fairclough Building Ltd* (1994) concerns category (1), indicating contributory negligence is inapplicable, although a sub-contractor in this case was able to argue successfully that he was liable in tort as well as contract, and for that reason was able to obtain a reduction in damages of 50% from another sub-contractor, in accordance with the *Vesta* v. *Butcher* category (3) position.

1.1 Negligence

This tort has been defined as 'the omission to do something which a reasonable man guided by those considerations which ordinarily regulate the conduct of human affairs would do or doing something which a prudent and reasonable man would not do' (see *Blyth* v. *Birmingham Waterworks Co* (1856)).

The following must exist in order to succeed in an action for negligence:

(1) The defendant must owe a duty of care to the plaintiff.
(2) There must be a breach of that duty.
(3) The plaintiff must sustain injury or damage as a result.

The duty of care

Only a knowledge of the law will indicate when a duty of care is owed. The contractor owes a duty of care to those people with whom he might come into contact in carrying out his work (and their property) but, for example, there is no duty to rescue a child drowning in a pond. Legal liabilities are not necessarily moral ones.

The most important test as to whether a duty of care exists was propounded by Lord Atkin in *Donoghue* v. *Stevenson* (1932) when he said:

> 'The rule that you are to love your neighbour becomes in law, you must not injure your neighbour; and the lawyer's question, Who is my neighbour? receives a restricted reply. You must take reasonable care to avoid acts or omissions which you can reasonably foresee would be likely to injure your neighbour. Who, then, in law is my neighbour? The answer seems to be – persons who are so closely and directly affected by my act that I ought reasonably to have them in contemplation as being so affected when I am directing my mind to the acts or omissions which are called in question.'

This is the famous case of an alleged snail in a bottle of ginger-beer causing shock and illness to the friend of the purchaser of the bottle, when she drank the contents. This is a test of reasonable foreseeability. Thus the decorator who leaves the premises in which he is working unattended and the door unlatched is liable for the resulting loss by theft because it was foreseeable (see *Stansbie* v. *Troman* (1948) and *Petrovitch* v. *Callinghams* (1969).

There are a number of circumstances where, apart from the moral aspect mentioned earlier, no duty of care is owed in spite of damage being foreseeable. In *Langbrook Properties Ltd* v. *Surrey County Council* (1970) it was held that the owner of land owes no duty of care to his neighbour for the results of his abstraction of water percolating underground in undefined channels. This decision follows the old case of *Bradford Corporation* v. *Pickles* (1895), where even a malicious act in relation to such abstraction carried no liability. Therefore it would be illogical to impose a duty for careless acts of this nature.

In *East Suffolk Rivers Catchment Board* v. *Kent* (1941) the Board had power to repair sea walls but were under no duty to do so. A very high tide flooded the respondent's land. The Board, requested to take action, did so with an allocation of manpower and resources which was hopelessly inadequate and which resulted in the respondent's land being flooded for much longer than it need have been. The House of Lords decided the Board were not liable because, as a matter of policy and exercising their discretion, they had decided not to devote a great deal of resources to the work.

Economic loss

When a contractor constructs a building defectively, three situations might arise as a result of the defect in the building (latent or patent), where there is no contractual relationship and the plaintiff is a third party dependent upon tortious principles for recovery of damages:

- A third party can be injured.
- Third party property can be damaged, i.e. not the building being erected.
- Economic loss can be suffered by the building owner or another third party.

There is no doubt that in the case of the first two, the contractor will be liable by the reasonable foreseeability rule enunciated in *Donoghue* v. *Stevenson* (1932). In the cases of *Clayton* v. *Woodman & Son (Builders) Ltd* (1962) and *Clay* v. *Crump* (1965) it was the builder's employee who was injured but the contractor would be liable to any other person who visits or passes by the buildings and who is injured or whose property is damaged in this way.

On the other hand, the courts were always reluctant to recognise a duty in negligence to avoid causing pure economic loss without any attendant physical damage to property. There is the principle that, if the result is foreseeable, negligent misstatement can give rise to a claim for damages for purely economic loss (*Hedley Byrne & Co Ltd* v. *Heller & Partners Ltd* (1964). However, in *Electrochrome Ltd* v. *Welsh Plastics Ltd* (1968) the defendant negligently damaged a fire hydrant causing the water supply to the plaintiff's factory to be cut off and it was held that while a duty of care was owed to the owners for the physical damage to their hydrant, there was no duty owed to the plaintiff, who was not the owner of the hydrant, and had only sustained loss of profit which was purely an economic loss. Even where physical damage to property was sustained, only the economic loss flowing directly from that physical damage was allowed in damages. This resulted in a very arbitrary line being drawn for economic loss claims.

Thus in *Spartan Steel & Alloys Ltd* v. *Martin & Co. (Contractors) Ltd* (1972) the all too frequent situation arose of a contractor damaging a cable in the ground and cutting off the power to a nearby factory occupied by the plaintiffs. They suffered some loss in value of the molten metal in a furnace plus the profit arising from that melt. They also claimed for the loss of profit in respect of melts which were delayed. This last item was disallowed as economic loss which did not flow directly from the physical damage, an example of judicial policy-making and of the courts drawing a line to avoid the risk of liability for an indeterminate amount of loss suffered by an indeterminate number of complainants for an indeterminate period of time, as the American judge, Cardozo CJ, put it in *Ultramares Corporation* v. *Touche* (1931).

The case of *Anns* v. *London Borough of Merton* (1977) and *Junior Books Ltd* v. *Veitchi Co Ltd* (1982) seemed to extend the duty arising from negligent action or inaction or breach of statutory duty to allow recovery of economic loss but this trend has been decisively reversed. *Junior Books* is limited to its facts and *Anns* has been overruled.

Landmark decisions in *D & F Estates Ltd* v. *Church Commissioners for England* (1989), in *Murphy* v. *Brentwood DC* (1991) and in *Department of the Environment* v. *Thomas Bates & Son* (1991) established that in construction cases there is no liability for pure economic loss which follows as a result of negligent design, supervision, construction or inspection. Recovery of any losses, in the absence of damage to property (other than the building constructed) or injury, is ruled out. Even if there is damage to other property, the courts might not recognise a duty of care to avoid all forms of economic loss, as the decision in the *Spartan Steel* case illustrates.

Lord Bridge in *Murphy* v. *Brentwood DC* (1991) put the position as follows:

> 'But if the defect becomes apparent before any injury or damage has been caused, the loss sustained by the building owner is purely economic. If the defect can be repaired at economic cost, that is the measure of the loss. If the building cannot be repaired, it may have to be abandoned as unfit for occupation and therefore valueless. These economic losses are recoverable if they flow from breach of a relevant contractual duty, but, here again in the absence of a special relationship of proximity, they are not recoverable in tort.'

In *Anns* v. *London Borough of Merton* (1977) a local authority failed to inspect or inspect properly (under their statutory powers) the foundations of dwellings in course of construction, which resulted in failure to notice faults leading to structural defects in those dwellings, and they were held liable for the loss occurring in tort. However, the House of Lords in *Murphy* v. *Brentwood DC* (1990) decided that it was illogical to hold a local authority liable in a situation where the builder responsible for the defective foundations was not liable. The implication of *Anns* was that a builder owes a duty of care to subsequent purchasers of his building for defects in it without any contractual nexus between them, but the House of Lords in *Murphy* saw no reason to distinguish between the sale of goods and the sale of buildings. Tortious liability should be limited to damage to property or injury to persons caused by a defect in goods or buildings but not for the inherent defects. The damage to the buildings gave rise to economic loss not physical damage. The cost of remedying a defect in property could not be described as physical damage to property. No property had been damaged, the building was merely defective. Thus the loss was economic.

Relationship of proximity

The reference by Lord Bridge to a special relationship of proximity stems from the case of *Hedley Byrne & Co Ltd* v. *Heller & Partners Ltd* (1964) allowing claims for economic loss flowing from negligent advice and the construction industry case of *Junior Books Ltd* v. *Veitchi Co Ltd* (1982). Apart from the *Hedley Byrne* case, until *Junior Books* it was believed that for an economic loss claim to succeed it had to be related to personal injury or damage to property (see the case of *Spartan Steel* above). But *Junior Books Ltd* v. *Veitchi Ltd* (1982) showed this to be untrue, at least as regards the position between employer (the person commissioning the work) and a nominated sub-contractor, who was liable in tort for defective flooring where personal injury and damage to property were not involved. The decision depended on the proximity of the relationship between the nominated sub-contractor and the employer (who were not in a contractual relationship), which was held to be close enough to give rise to a duty of care to avoid defects in the work itself. This duty extended to include the avoidance of causing economic loss consequential on the work being defective.

In *Junior Books Ltd* v. *Veitchi Co Ltd* (1982) the only property affected was the actual property being constructed (in any event the production of a defective product is not legally regarded in the same light as damaging property), and it was doubted whether valid economic loss claims had to be based on physical damage to property.

While *Junior Books* has been restricted, limited to its own particular facts, and even isolated, it has never been overruled. Thus in *Greater Nottingham Co-op* v. *Cementation Piling & Foundations Ltd* (1989) it was held that the existence of a contract between the employer and the nominated sub-contractor, which did not help the employer in the particular circumstances as it did not deal with the manner in which the work was executed, nevertheless prevented the decision in *Junior Books* applying. In fact it was said by Lord Bridge in *D & F Estates Ltd* v. *Church Commissioners of England* (1988) about *Junior Books*:

'The consensus of judicial opinion, with which I concur, seems to be that the decision of the majority is so far dependent upon the unique, albeit non-contractual relationship between the pursuer and the defender in that case and the unique scope of the duty of care owed by the defender to the pursuer arising from that relationship that the decision cannot be regarded as laying down any principle of general application in the law of tort or delict.'

While it is clear that *Junior Books* lays down no principle of general application, it seems to lay down principles concerning the employer and nominated sub-contractor relationship where there is no collateral warranty agreement. It has been said that the apparent assumption in the above quotation that the relationship between the defender and the pursuer in *Junior Books* was unique is surely misplaced as there can hardly be a more common relationship than that between an employer and a nominated sub-contractor. The point is that the *Junior Books* decision is still alive but limited to its own particular facts and the absence of a collateral warranty which helps the employer.

The complex structure theory

When, in *D & F Estates*, the House of Lords restricted the tort liability of contractors to those defects which caused personal injury or damage to other property (meaning property other than the property the builder was erecting) it was not decided whether 'other property' might be taken to include other parts of the building being erected. Thus the owner having the work done might be able to claim that the defective foundations have caused damage to the walls of the building they supported. In *Murphy* this theory was rejected. Consequently 'other property' means only such things as a car or furniture damaged by a falling roof. The structure of the building being erected is not included.

Although rejecting the complex structure theory the House of Lords in *Murphy* said that a sub-contractor, employed to construct part of a building (e.g. heating installation) might be liable if his negligence caused damage to another part of the building. For example, if a boiler exploded and damaged the rest of the structure. This presumably follows the normal *Donoghue* v. *Stevenson* principle.

Developments after *Murphy*

The decision in *Department of Environment* v.*Thomas Bates & Son* (1991) inevitably followed *Murphy*. This case concerned an eleven storey block which the Department leased in 1971. In 1979 it was discovered that the support columns had been formed from an incorrectly mixed concrete and the columns would be incapable of supporting the original design load of the building, though there has been no actual damage since the full load had never been imposed. The Department had the necessary repairs carried out and sued the builders for recovery of the costs. The action was based on negligence since there was no contract between the Department as lessee and the builder. The conclusion in the House of Lords was that, as the House had held in *Murphy* that *Anns* v. *London Borough of Merton* (1978) had been wrongly decided, and as the loss suffered by the Department had been entirely economic, the appeal should be dismissed. There was no sound basis in principle for holding the builder liable for the pure economic loss suffered by a purchaser who discovered the defect and was required to expend money in order to make the building safe and suitable for its intended purpose.

In *Preston and Another* v. *Torfaen Borough Council and Another* (1993) the council before building a housing estate engaged soil engineers and, on the basis of their report, started work. The plaintiffs were originally tenants of the council but later purchased a flat on the estate under a scheme open to them. Subsidence occurred and the plaintiffs sued the council and the engineers. The claim against the council was abandoned in view of the decision in *Murphy*, but continued against the engineers. The Court of appeal held that there was insufficient proximity between the plaintiffs and the engineers to support a duty of care for economic loss under the *Hedley Byrne* decision. At the time of the alleged negligence, the plaintiffs would have been unknown to the engineers, and the engineers could not owe a duty of care for economic loss to a member of a potential class of purchasers.

However, as one door closes another opens as it has to be recorded that in 1994 and 1995 two construction cases have shown that there is no sustainable distinction between the making of statements and other exercises in duty causing economic loss. Thus in *Conway* v. *Crowe Kelsey & Partner and Anchor Foundations Ltd* (1994) a design was considered a 'statement' for the purposes of liability on the *Hedley Byrne* v. *Heller* principle. In 1995 the Court of Appeal in *Barclays Bank plc* v. *Fairclough Building Ltd* (1994) decided that in the case of a sub-subcontractor hosing down an asbestos roof (there was no defect in the product of the works, i.e. the cleanliness of the roof) there was liability for economic loss by contaminating the plaintiff's other property caused by dangerous asbestos fibres which loss arose from defects in the way in which the cleaning was carried out. This means that these circumstances are indistinguishable from the case of a professional man making a statement. Undoubtedly liability pursuant to *Hedley Byrne* has expanded.

The standard of care

Consequently, only by knowing the law is one able to indicate when the duty of care is owed. However, when the duty of care is owed, the standard is an objective one, i.e. how a hypothetical reasonable man would behave in a given set of circumstances; not what is reasonable for that particular defendant to do or not to do in a particular case. Nevertheless, the amount of care can vary with the circumstances. Thus the contractor in occupation of property owes the same standard of care to both adults and children, but he must exercise more care if the premises are used by children rather than only by adults, e.g. in the extension of the wing of a school when the main building is still used by the pupils.

Another illustration is the case of *Haley* v. *London Electricity Board* (1964), where it was decided that in view of the number of blind persons in the population (1 in 500), contractors who opened a hole in a town highway must take reasonable care to prevent a blind person (using reasonable care for his own safety) from falling down the hole. This obviously imposes a greater amount of care upon contractors and also upon local authorities.

The burden of proof

As indicated earlier in civil cases the burden of proof is upon the plaintiff to prove his case on the facts pleaded by him and not admitted by the defendant. The standard is proof on a balance of probabilities. This onus of proof applies to negligence but, in certain cases where the maxim *res ipsa loquitur* (the thing speaks for itself) applies on proof of the occurrence, the onus is not (as at one time thought) transferred to the defendant to disprove negligence (see *Ng Chun Pui and Others* v. *Lee Chuen Tat and Another* (1988). Rather the doctrine is now stated as establishing a prima facie case for the plaintiff.

Three requirements are necessary before the rule can apply:

- The thing causing the accident must be under the defendant's exclusive control.
- The accident is such as, in the ordinary course of things, does not happen if those who have the management use proper care.
- There must be no explanation of the accident. This means that if proof of the relevant facts is available, the question ceases to be whether they speak for themselves. The only question is whether, on the established facts, negligence can be inferred or not.

If the defendant contractors were demolishing a building adjacent to the highway and the plaintiff, a passer-by was injured by falling masonry, assuming there was no evidence as to how the masonry came to fall, *res ipsa loquitur* would apply. In *Walsh* v. *Holst* (1958) on similar facts the defendants escaped liability as they were able to show that all proper and reasonable precautions had been taken.

1.2 Nuisance

One of the earliest definitions of nuisance is 'the wrong done to a man by unlawfully disturbing him in the enjoyment of his property [a private nuisance], or, in some cases, in the exercise of a common right [a public nuisance]'. The insertions within the brackets have been made by the author.

Public nuisance

This is a crime and only falls within the law of torts when it causes a person special damage above that caused to the community in general. Thus the builder who leaves a pile of sand on the highway and fails to illuminate it at night would be guilty of a public nuisance because of the general inconvenience to the public who have to avoid the obstruction. It is for the police to take action in the case of a public nuisance; if a car driver collides with the unlit obstruction and sustains injury, he would also have a right of action against the builder as he suffered an injury over and above that caused to the generality of the public. Another example is the creation of dust and smoke from building sites which obscure the view of drivers using the highway. If an accident is caused by such dust or smoke, a public nuisance is present allowing an action in tort by those injured.

Public nuisance also includes the creation of dangers upon or near the highway and liability for the disrepair of premises abutting the highway. There is some tolerance in respect of building works adjoining the highway, such as hoardings erected outside the area of building activities, in that they encroach upon the highway but are for the public's protection and so acceptable. Moreover, there is also statutory authority for such creations, unless negligence has taken place in the erection of the hoardings.

Private nuisance

This was defined in *Cunard* v. *Antifyre* (1932) as 'an interference for a substantial length of time by owners or occupiers of property with the use or enjoyment of neighbouring property'.

Private nuisances are of two kinds:

- Wrongful disturbance of rights attaching to land. The main rights, interference with which may be actionable, are rights to light and air, rights to the support of land and buildings thereon and rights in respect of water and rivers.
- The act of wrongfully causing or allowing the escape of noxious things, such as smoke, smells, noise, gas, vibration, damp and tree roots, into another person's property so as to interfere with his health, comfort, convenience, or enjoyment of, or cause damage to, his property.

The following points are noteworthy:

(1) The common feature of both public and private nuisance is unlawful interference. This interference must be real, not fanciful and must normally involve something continuous. This does not mean that the damage must be continuous as there are cases of successful actions in nuisance involving a single occurrence of damage. However a state of affairs must exist; the term 'nuisance' implies some continuity. There must, moreover, be some resulting damage, discomfort or inconvenience to satisfy the requirement that some damage, etc. be caused. In *Andreae* v. *Selfridge & Co Ltd* (1938), where a hotel proprietor claimed for loss of custom through building operations, the Court of Appeal reversed an award of damages amounting to the full extent

of the loss of custom, holding that some of the interference was reasonable yet might have resulted in loss of custom. Consequently the court calculated the proportion of the business loss equivalent to the excess of noise and dust which alone was actionable.

(2) Apparently in private nuisance no English case has been founded solely upon bodily injury. Private nuisance is limited to inconvenience or damage caused in respect of the enjoyment of land, and if personal injury is suffered without affecting the occupier's enjoyment of land, the action will probably be for negligence or trespass. However, in principle there seems no reason why a person who suffers an invasion of some proprietary or other interest in land should not be able to claim for a resulting personal injury.

(3) Liability in nuisance can apply without a breach of any duty of care, e.g. to avoid the emission of dust in building activities. This distinguishes this tort from that of negligence. Another example which owners of buildings, having alterations carried out, seem to be guilty of from time to time is the removal of support to neighbouring land and the buildings thereon.

(4) An action may lie in both nuisance and negligence. In *Gold* v. *Patman & Fotheringham Ltd* (1958), the defendants undertook work on Gold's property. Piling operations took place without negligence on the contractor's part, but third party property sustained damage due to weakening of support. The third party property owner succeeded in an action in nuisance against Gold. Had the contractors been negligent, there would have been liability on the part of Gold because he would have been vicariously responsible for the contractor's negligence in not taking precautions as well as directly for his own nuisance. The interference with the right of support is one of those circumstances when a principal is liable for his independent contractor's tort, i.e. nuisance in this case.

(5) As the interest protected in private nuisance is the proprietary interest in land, only someone in possession of land or who has a proprietary interest can sue, i.e. the owners or tenant but not someone merely using the land. A reversioner (someone with a future interest in land) can only sue if his reversion is affected, which usually means serious and irreparable damage to the land. As already mentioned, any person suffering special damage (see earlier for the meaning of this term) can sue in public nuisance. This could be the occupier of adjacent land or a user of the highway.

(6) In the case of a nuisance created by a positive act of misfeasance (the improper performance of a lawful act), the creator of the nuisance is liable for it and its continuance even when he has lost the ability to abate it. Consequently, when a building was erected so that it obstructed ancient lights (the right of access of light to any building actually enjoyed for twenty years without interruption), the builder was held liable in nuisance even though he had assigned the lease and was therefore unable to discontinue the nuisance. See *Roswell* v. *Prior* (1701).

(7) An occupier must refrain from creating or continuing a nuisance, or prevent one coming into existence, whether caused by others or by natural causes. This liability arises from the fact that an occupier has full knowledge of the activities on and control over the land occupied. In *Matania* v. *National Provincial Bank* (1936) the occupier of the first floor of a building was liable to the occupier of the higher floors for a nuisance by dust and noise created by his independent contractor. The latter would also be liable as the creator of the nuisance.

(8) A landlord or owner of property but out of occupation may be liable if he:

● has authorized the nuisance;
● knew or ought to have known of the nuisance before letting and did not stipulate for the tenant to put it right;
● has reserved the right to enter and repair or has an implied right to do so;
● is under an express covenant to repair.

1.3 Absolute or strict liability

Probably the word 'strict' is the better term as 'absolute' implies that the defendant is always liable and this is not so. There are defences available, and the word 'strict' more aptly describes the situation. This is that there are certain liabilities imposed by law making the defendant liable even though he has exercised reasonable care and did not intend the injury or damage.

Rylands v. Fletcher

The usual example under this heading is the rule in *Rylands* v. *Fletcher* (1868) that the person who, for his own purposes, brings on to his land and collects and keeps there anything likely to do mischief if it escapes must control it at his peril, and, if he does not do so, is prima facie answerable for all the damage which is the natural consequence of its escape. Therefore there is a strict liability not requiring proof of negligence.

In *Cambridge Water Company* v. *Eastern Counties Leather* (1994) the House of Lords decided that foreseeability of harm or injury of the type complained of was a prerequisite of the recovery of damages under the rule in *Rylands* v. *Fletcher*. The defendants operated a tanning factory. Regular but small spillages of a solvent seeped through the factory floor and into the soil, from where it contaminated the plaintiff's borehole over a mile away, by means of percolating water. The plaintiff was required by EC legislation to sink a new borehole, and claimed the cost of this from the defendants in negligence, nuisance and under *Rylands* v. *Fletcher*.

The actions in negligence and nuisance were dismissed by the trial judge on the ground that the loss was not foreseeable. He also rejected the *Rylands* v. *Fletcher* claim on the basis that the storage and use of the solvent was a natural use of the defendant's land, as it was used for industrial purposes. On the plaintiff's appeal against the dismissal of this third cause of action the Court of Appeal declined to determine it on the basis of the rule in *Rylands* v. *Fletcher*, holding instead that there was strict liability in nuisance under *Ballard* v. *Tomlinson* (1885), in which liability had been imposed in similar circumstances.

On appeal to the House of Lords it was held that foreseeability of harm of the relevant type was a prerequisite of the recovery of damages both in nuisance and under *Rylands* v. *Fletcher*. Consequently, although the House also decided that the storage of the solvent constituted non-natural use of the defendant's land, since the plaintiff could not establish that the pollution was in the circumstances foreseeable, the action failed.

Therefore, it appears that the defendant's strict liability under the rule merely means that the defendant cannot contest liability by showing that he had taken all reasonable steps to prevent the escape.

Defences to Rylands *v.* Fletcher

There are the following defences:

- The escape was due to the plaintiff's default.
- It was due to an act of God.
- The incident results from natural use of the land.
- It was done with the plaintiff's consent.
- There was statutory authority to carry out the activity.
- It was the act of a stranger. The word 'stranger' cannot include a servant or contractor, for whose acts the defendant would be responsible, but for example, there would be no liability under the rule of the act of a trespasser.

The decision of the Court of Appeal in *H. & N. Emanuel* v. *Greater London Council and Another* (1971) illustrates *inter alia* the rule and includes a discussion of the meaning of 'stranger' in the last defence mentioned above. The Council required the Ministry of Housing to demolish two prefabricated houses on their land. The Ministry of Works at the request of the Ministry of Housing sold the houses to a contractor, Mr King, for removal under a contract forbidding King to light any fires, which the Ministry knew King had done previously despite prohibitions.

King started the work of demolition and failed to control a bonfire he lit, and the plaintiff's premises were damaged by fire. The Court unanimously held the Council liable. Two of the judges seemed to consider that the judge in the court of first instance was right in finding the Council liable under the *Rylands* v. *Fletcher* rule, but in any event decided that King was not a 'stranger' and that the Council were 'occupiers' of the premises because they had a sufficient degree of control over the activities of persons thereon. The Council could reasonably have anticipated that King's men would light a fire and ought to have taken more effective steps to prevent them. Accordingly the Council were themselves negligent, apart from any liability under the *Rylands* v. *Fletcher* rule.

Fire

It is important to make two points. In the first place, fire is a special case. If it arises from a dangerous thing being brought upon the land and if the fire escapes to damage third party property there is a strict liability; the Fires Prevention (Metropolis) Act 1774, which provides that there is no liability for a fire starting accidentally, does not apply. Thus in *Balfour* v. *Barty-King and Others* (1957), a pipe in the loft of the defendant's premises became frozen. Workmen in the employment of the third party to the action used a blow-lamp and the pipe lagging caught fire. This method of unfreezing the pipe was sufficient to operate the rule in *Rylands* v. *Fletcher* so far as the plaintiff's claim against the defendants was concerned. The third party contractor was also held liable to reimburse the defendants in respect of the award against them and pay for the damage to the defendant's property.

The second point is that where a contractor is employed to do work which involves a strict liability such as that concerning the *Rylands* v. *Fletcher* rule, the principal becomes vicariously responsible for the contractor's acts (see below).

Incidentally, under a fire policy (which includes a fire and special perils policy or a contractors' all risk policy), the insurer only agrees to pay the value of the property up to the policy limit; he does not pay for the consequential loss caused by the insured being deprived of the use of the insured property until it is repaired or rebuilt. To recover such a loss, the owner requires a consequential loss policy (in the case of buildings in course of erection it is called an advance profits policy, see Chapter 7) or he is compelled to bring an action in tort.

Contract/tort

Alternatively, if the contractor is responsible in contract, e.g. clause 22A of the JCT Form, then the employer (owner) can hold the contractor responsible under contract, but whether this clause includes consequential loss (apart from physical loss or damage) is doubtful (see Chapter 11). If, as in the *Emanuel* case, the claimant is a third party, he would not be under contract to the contractor and would have to bring his action in tort.

Vibrations

Another example of the operation of strict liability occurred in *Hoare* v. *McAlpine* (1923), where a contractor drove a number of piles into the ground and, due to the resulting vibrations, caused damage to a house belonging to the plaintiff. The contractor was held liable to the plaintiff without proof of negligence. It should also be noted that vibrations could constitute a nuisance. However, it may be a defence to nuisance that the property damaged is particularly sensitive, e.g. that the house in *Hoare's* type of case is abnormally unstable, whereas liability under the *Rylands* v. *Fletcher* rule is strict. However, in *Barrette* v. *Franki Compressed Pile Co of Canada Ltd* (1955), pile-driving vibrations were not held to be within *Rylands* v. *Fletcher* but to be a nuisance.

Vicarious liability

Another illustration of strict liability occurs where one person is vicariously responsible for the acts or omissions of another. This occurs in the following relationships (the first-named being responsible for the acts of the second-named in each case):

● master and servant, usually in all circumstances;
● principal and independent contractor, in certain circumstances only;
● principal and agent, in particular situations only.

1.4 Trespass

This has been defined as an unlawful act committed with force and violence on the person, property or right of another. The 'violence' may be only implied, e.g. mere wrongful entry on to the plaintiff's land or even to his air space by a tower crane. In *Konskier* v. *Goodman* (1928) a builder was allowed to leave rubbish while demolishing part of a building. During demolition Konskier became tenant of the premises. After completion of the work the builder did not remove the rubbish and it was held that the tenant could recover in trespass. The injury must be 'direct', which means that it must not be consequential. Thus, if a builder throws a brick into the road injuring a passer-by, it is a direct injury, but if the brick thrown into the road is later the cause of injury to a passer-by who falls over it, the injury is an indirect one.

For all practical purposes there is no significant difference between unintentional trespass and the tort of negligence causing injury. See *Fowler* v. *Lanning* (1959) and *Letang* v. *Cooper* (1965). Incidentally, liability policies do not cover intentional injury or damage, although this would be a trespass. The usual example is the builder who digs a trench across a field of crops damaging them. He is guilty of trespass and probably other wrongs, but the public liability policy would not cover such damage because it is not accidental; it is inevitable, which is a standard exclusion of this policy.

Looking at trespass from the viewpoint of the occupier of the land upon which the trespass takes place, it should be appreciated that, whereas at one time it was said that no duty of care was owed to a trespasser (although traps should not be set), the modern view is that the occupier is expected to act with ordinary humanity. This can result in a duty to prevent the trespasser getting to the property, particularly where it is dangerous to children. See the case of *Pannett* v. *McGuinness & Co Ltd* (1972), mentioned later when discussing the Occupiers' Liability Acts 1957 and 1984.

1.5 Liability under contract

The purpose of this book is to consider the various forms of contract issued by the Joint Contracts Tribunal from an insurance point of view. As will be seen later in the book, the Standard Form of Building Contract, like most construction contracts, increases the contractor's liability beyond that at common law. However, under this heading it is intended briefly to consider some of the rules concerning the law of contract.

A contract has been defined by Treital in his book *An Outline of the Law of Contract* as: 'an agreement which is either enforced by law or recognised by law as affecting the legal rights or duties of the parties'.

The essentials affecting a contract on standard terms are:

- An unconditional offer and an unqualified acceptance.
- Unless the contract is under seal, e.g. a performance bond, consideration must be given by the party suing. Consideration was defined in *Currie* v. *Misa* (1875) as some right, interest, profit or benefit accruing to the one party or some forbearance, detriment, loss or responsibility given, suffered or undertaken by the other.
- There must be an intention to create a legal relationship. Normally commercial agreements are intended to create legal obligations.

Incidentally, public liability policies catering for the building and allied trades amend the basic exclusion (of liability assumed under agreements except where the liability would have attached even in the absence of agreement) to cover contractual liability.

In the JCT Standard Form of Building Contract, the wording has been agreed by the representative bodies of both parties to the contract. The general rule is that people are free to enter into contracts or not as they please. In the case of the construction industry the contractor can ask for any amendment, but this should only be done after very careful consideration as often one clause depends upon another and alterations can make nonsense of other parts of the contract.

1.6 Breach of statutory duty

Certain Acts of Parliament create statutory liabilities. The following are among those most likely to be involved in claims concerning the construction industry.

(1) The Fatal Accidents Act 1976

This Act consolidates the previous Fatal Accidents Acts and gives a right of action to the dependants of those killed in accidents to the extent of their pecuniary loss, provided that the deceased would have had such a right had he lived.

(2) The Law Reform (Miscellaneous Provisions) Act 1934

This has the main effect of providing that, on the death of any person, all causes of action subsisting against or vesting in him at the time of his death shall survive against or for the benefit of his estate.

(3) The Administration of Justice Act 1982

This provides, *inter alia*, for the inclusion in a claim under the Fatal Accidents Act of damages of a fixed sum for the close relatives of the deceased for their bereavement.

(4) The Factories Act 1961

This is another consolidating Act which is similar to other enactments relating to the safety, health and welfare of employees. The Factories Act itself is not applicable to new building work, but it will apply to the occupier of a factory within which building work is being done. As the employer (principal) having the work done, he will be liable for injuries due to a breach of the Act, and this liability is not limited to the principal's own employees.

(5) The Health and Safety at Work etc. Act 1974

The 1974 Act established the Health and Safety Commission as the overall authority to achieve the purposes of the Act, and has brought together within the Health and Safety Executive five inspectorates, including one for factories, to enforce the law. It imposed new comprehensive statutory obligations on employers and others, and presented new duties and rights to employees (see sections 2–9). Breaches of these sections by themselves carry no liability to a civil action, only to a criminal prosecution. However, it should be noted that while the Health and Safety at Work etc. Act 1974 is backed by criminal sanctions, under section 47(2) any breach of a regulation made under the Act will confer a right in civil law unless that regulation provides otherwise.

Construction industry regulations

The intention was that the 1974 Act should supersede all existing safety legislation, replacing it gradually with regulations and Codes of Practice which would, according to the wording, give a civil right of action. However, at the time of writing, everyday operations in the construction industry are governed by the following regulations, published under the Factories Act then in force and the 1974 Act:

- The Construction (General Provisions) Regulations 1961
- The Construction (Lifting Operations) Regulations 1961
- The Construction (Working Places) Regulations 1966
- The Construction (Health and Welfare) Regulations 1966
- The Construction (Head Protection) Regulations 1989

During the last twenty years there have been a myriad of regulations passed to protect the worker's health and safety, The Management of Health and Safety at Work Regulations 1992 are probably the most far-reaching that have been introduced since the Health and Safety at Work etc. Act 1974. These regulations require specific action to be taken to show that correct procedures have been taken to meet the requirements of section 2 of the 1974 Act, which mainly sets out the common law duty of care of employers. Employers must identify health and safety problems and then produce the means of overcoming them.

Employers (masters in the master-and-servant relationship) are responsible under the civil as well as the criminal law to their own employees for breaches of regulations, and contractors and users of plant or equipment also owe certain duties under them. Employees and self-employed persons are also required to comply with these regulations.

The Construction (Design and Management) Regulations 1994

These became operative on 31 March 1995. They deal with the responsibilities of the client, the designer, the planning supervisor, the principal contractor and other contractors, so as to impose requirements and prohibitions with respect to the design and management of construction sites.

Breach of a duty imposed by these Regulations, other than those imposed by Regulation 10 and Regulation 16(1)(c), do not confer a right of action in any civil proceedings. Regulation 10 reads 'Every client shall ensure, so far as is reasonably practicable, that the construction phase of any project does not start unless a Health and Safety plan complying with Regulation 15(4) has been prepared in respect of that project'. Regulation 16(1)(c) reads 'The principal contractor appointed for any project shall take reasonable steps to ensure that only authorised persons are allowed into any premises as part of premises where construction work is being carried out'.

(6) The Occupiers' Liability Act 1957

This Act reduces the categories of persons entering premises to visitors and trespassers. It does not provide any protection for trespassers. The duty owed to a trespasser was mentioned when discussing the tort of trespass, and that duty is to take such steps as common sense or common humanity dictate to reduce or avoid the dangers and which are reasonably practicable to the particular occupier concerned. This was laid down in *Herrington* v. *British Railways Board* (1972), the case of a child injured on an electric railway line. An example from the construction industry is *Pannett* v. *McGuinness & Co Ltd* (1972) where the Court of Appeal held that an occupier/contractor was not absolved by previous warnings to a child trespassing on a demolition site which was left unprotected at a time when the site was at its most dangerous and alluring.

Under the 1957 Act the common duty of care is owed to all visitors. This is 'a duty to take such care as in all the circumstances of the case is reasonable to see that the visitor will be reasonably safe in using the premises for the purposes for which he is invited or permitted by the occupier to be there'. In contracts for new building work, the site and works are occupied by the contractor until they are handed over to the employer (owner), and thus the contractor is the occupier within the meaning of the Act.

However, in *Wheat* v. *Lacon* (1966) the manager of a public house and the brewery company who owned the premises were both held to be occupiers and so they each owed the common duty of care to their visitor. In similar circumstances an employer (owner) and contractor can be occupiers of the same premises under the Occupiers' Liability Act, as in *AMF International* v. *Magnet Bowling Ltd* (1968), where both the main contractor and the employer were held to be occupiers with respect to a direct specialist contractor's claim. A sub-contractor can also be an occupier of the whole or part of a site.

(7) The Occupiers' Liability Act 1984

This is concerned with the liability to entrants other than 'visitors', principally trespassers. Thus an occupier owes a duty to another only if:

- he is aware of the danger or reasonably believes that it exists;
- he knows or reasonably believes that the other is in (or may come into) the vicinity of the danger; and
- the risk is one against which ... he may reasonably be expected to offer the other some protection.

This seems to be the decision in *Herrington* put into statutory form.

(8) The Building Act 1984

Part I of the Building Act is concerned with building regulations; Part II deals with the system of private certification, i.e. the supervision of building work etc. otherwise than by local authorities; and Part III deals with miscellaneous but related matters.

A building is defined in the Act as:

'... any permanent or temporary building and, unless the context otherwise requires, it includes any other structure or erection of whatever kind (whether permanent or temporary).'

'Structure or erection' includes a vehicle, vessel, hovercraft, aircraft or other movable object of any kind in such circumstances as may be prescribed by the Secretary of State. This wide definition is fortunately not followed in the definition of 'building' in the Building Regulations 1985 (see below). However, a wide definition is necessary for the general purposes of the Act, e.g. in connection with local authorities' power to deal with dangerous structures. The Building Regulations define a building as 'any permanent or temporary building but not any kind of structure or erection'. This definition includes a part of a building.

The powers and duties of a local authority in regard to the passing or rejection of plans are stated in section 16 of the Act. Also the Building Regulations contain provisions as to notices and forms. However, there are provisions by which Building Regulation compliance or approval may be obtained other than by a decision of the local authority. The main alternative appears under section 17 of the Act whereby 'approved persons' may be designated to give a certificate of compliance with Building Regulations. This person must also provide evidence of insurance arrangements complying with regulations.

Section 38 provides that breach of a duty imposed by the Building Regulations is to be actionable in civil proceedings unless the Regulations provide otherwise, so far as it caused damage. 'Damage' is defined as including the death of, or injury to, any person. The Regulations may provide prescribed defences to such an action. The provision does not prejudice any right of action at common law, but has not yet been activated.

(9) The Building Regulations 1985

There have been amendments to these Regulations from time to time and as recently as 1994. The Regulations apply to England and Wales but not to Scotland and Northern Ireland, where a separate system applies. They set reasonable standards of health and safety, but do not stipulate in detail the technical means of achieving these. Technical details are found in fourteen Approved Documents plus certain other non-statutory guides.

Civil liability can occur under the Regulations on the part of contractors both contractually by express provisions under the contract: clause 6.1.1 in the JCT form and through breach of statutory duty by failure to comply with the Regulations. Lord Wilberforce in *Anns* v. *London Borough of Merton* (1978) said:

> 'Since it is the duty of the builder, owner or not, to comply with the by-laws, I would be of the opinion that an action could be brought against him in effect for breach of statutory duty by any person for whose benefit or protection the by-law was made.'

While this remark concerning statutory duty is still valid the decision of the House on economic loss is no longer good law (see above).

(10) The Defective Premises Act 1972

This Act is best explained by reference to the relevant Law Commission report.

The Law Commission report on the Civil Liability of Vendors and Lessors for Defective Premises (Law Comm. No. 40) summarized four recommendations for amendment of the law. The first of these dealt with liability for defects of quality in newly built dwellings. The remainder dealt with liability for dangerous defects in premises of any description sold or let.

(a) Newly built dwellings

Those who build, or undertake work in connection with the provision of, new dwellinghouses (whether by new construction, enlargement or conversion) should be under a statutory duty owed to any person who acquires a proprietary interest in the property, as well as to the person (if any) for whom they have contracted to provide the dwelling, to see that the work which they take on is done in a workmanlike or professional manner (as the case may be), with proper materials and so that the dwelling will be fit for habitation. The same duty should be owed by those who in the course of trade or business or under statutory powers arrange for such work to be undertaken.

Section 1 of the 1972 Act achieves the object recommended, but it only operates if section 2 does not apply. That is, if an approved scheme is in operation in relation to the dwelling, then no action can be brought in respect of the duty in section 1 of the Act. Since 1979 the National House-Building Council (NHBC) has not applied for approval of its scheme which means that section 1 of the 1972 Act applies as well as any contractual remedy under the scheme. However, under section 1 there is a rather restrictive limitation as any cause of action accrues when the dwelling was completed, so time starts to run at a very early stage. Whereas in an action for negligence, by virtue of the Latent Damage Act 1986, the limitation period begins only when the plaintiff should reasonably have known of the relevant damage subject to a long-stop of fifteen years after which any action is invalid. It is not clear whether the 1986 Act applies to claims brought under the 1972 Act.

Section 1 contains a wide-ranging duty. It is placed on not only builders but on all those who take on work concerning the provision of *homes*. It should be noted that the other sections of the Act apply to all *premises*. Section 1 provides a specific form of liability. Most purchasers can avail themselves of the Act's protection, as illustrated by the case of *Andrews* v. *Schooling* (1991).

(b) Caveat emptor

The vendor's and lessor's immunity from liability for the consequences of his own negligent acts should be abolished; that is to say, caveat emptor ('let the buyer beware') should no longer provide a defence to a claim against a vendor or lessor which is founded upon his negligence.

Section 3 of the Act clearly attains this objective (as far as negligence in building or other work is concerned which is carried out on the land before the sale or letting).

(c) Liability for known defects

A person who sells or lets premises should be under the general duty of care in respect of defects which may result in injury to persons or damage to property and which are actually known to him at the date of the sale or letting. Section 3 also achieves this recommendation, subject to the statement in parentheses in the previous paragraph.

(d) Landlord's repairs

A landlord who is under a repairing obligation or has a right to do repairs to premises let should be under the general duty of care in relation to the risk of injury or damage arising from a failure to carry out that obligation or exercise that right with proper diligence; but a landlord should not, by reason of a right to enter and repair on the tenant's default, be liable to the tenant for injury suffered by the tenant's default.

Section 4 achieves the desired effect here.

(11) The Consumer Protection Act 1987

The two main purposes of this Act are to introduce the provisions of the EC Directive (85/374/EEC) concerning strict liability for defective products which result in personal injury or damage to private property (Part I) and to enforce a general standard of safety in the supply of products (Part II). Part II consolidates the Consumer Safety Act 1978 and the Consumer Safety (Amendment) Act 1986, and is part of the criminal law, although it may have some influence on civil liability. Part I came into force on 1 March 1988 although the Act received the Royal Assent on 15 May 1987. It applies only to products sold or supplied after the Act came into force. Part II came into force on 1 October 1987.

The purpose and construction of Part I section 1

Subsection (1) states that:

> '... this Part shall have effect for the purpose of making such provision as is necessary in order to comply with the Product Liability Directive and shall be construed accordingly.'

This is important as the Act does not always comply with the EC directive. Thus, for example, the definition of product includes goods (see below), which in turn is interpreted (in s. 45) as including 'things comprised in land by virtue of being attached to it', which in isolation would mean that a product includes a building. However, the directive states in Article 2 that 'product' means all movables, which would not include buildings, and therefore they are not covered by strict liability, although building components are. Furthermore the introductory paragraph of the interpretation section (s. 45) contains an exception to the definitions given in that section by stating 'except in so far as the context otherwise requires', and the context in section 1(1) requires the directive to be followed. Incidentally in section 4(1)(b) there is a defence if the person proceeded against can show that he did not supply the product. Thus the question arises as to whether a builder who builds a defective building actually supplies it within the meaning of the Act. This question is discussed further later in this chapter when considering section 46 on the meaning of 'supply'. In any event it is a pity that there should be even the slightest doubt at the outset of this legislation on a matter that is so important to the construction industry and those connected with it.

Product

Section 1(2) of the Consumer Protection Act 1987 defines product as any goods or electricity and (subject to subsection (3)) includes a product which is comprised in another product, whether by virtue of being a component part or raw material or otherwise. Apparently this definition includes gas, as goods include substances and substance according to section 45 means any substance whether in solid, liquid or gaseous form.

It has already been indicated that a product does not include a building, but clearly the components of a building are products. Therefore if a brick or tile is defective and as a result falls off the building injuring a passer-by then the manufacturer of that component is liable.

Producer

Subsection (2) defines producer, in relation to a product as:

'(a) the person who manufactured it;
(b) in the case of a substance which has not been manufactured but has been won or abstracted, the person who won or abstracted it;
(c) in the case of a product which has not been manufactured, won or abstracted but essential characteristics of which are attributable to an industrial or other process having been carried out (for example, in relation to agricultural produce), the person who carried out that process.'

The relevance of this definition of producer is seen under section 2(2) as the person liable for damage caused by a defective product is the producer including the importer in cases where the product has been imported into the EU and the own-brander where it has been own-branded. The builder will not be liable under section 2(2), unless he actually produces the product, and will be able to pass the liability on to whoever produced or imported it into the EU.

Supplier

Section 2(3) of the 1987 Act extends liability to suppliers of products in certain specially defined circumstances as follows:

> 'Subject as aforesaid, where any damage is caused wholly or partly by a defect in a product, any person who supplied the product (whether to the person who suffered the damage, to the producer of any product in which the product in question is comprised or to any other person) shall be liable for the damage if –
> (a) the person who suffered the damage requests the supplier to identify one or more of the persons (whether still in existence or not) to whom subsection (2) above applies in relation to the product;
> (b) that request is made within a reasonable period after the damage occurs and at a time when it is not reasonably practicable for the person making the request to identify all those persons; and
> (c) the supplier fails, within a reasonable period after receiving the request, either to comply with the request or to identify the person who supplied the product to him.'

Thus a builder will be a supplier of a product if he is unable or unwilling to identify the producer, under this subsection. Therefore it behoves the builder to keep a careful record of the origin of his materials. The contractor who simply installs doors in a house does not manufacture the product, but if he assembles the doors from component parts he will probably be considered the manufacturer of the doors. In the former case he should keep a record of his supplier.

Safety

This is the basis of the meaning of 'defect' (s.3). Whatever the kind of defect (production, marketing or design) the test is the same: safety in relation to a product, coupled with what persons generally are entitled to expect. Safety in this context includes safety concerning:

- products comprised in a product;
- risks of damage to property;
- risks of death or personal injury.

Therefore it is important to appreciate that a product is not defective within the meaning of the Act simply if it does not operate properly or has a shorter life than it should. 'Safety' is the key word.

Damage to the product itself

Section 5(2) states that a person is not liable for any defect in a product for the loss of or any damage to the product itself, or to any product which has been supplied with the defective product comprised in it.

Commercial property

Section 5(3) deals with the crucial distinction drawn between damage to private property which is covered and damage to commercial property which is not. In the latter event the claim would have to be in contract or negligence. There are said to be two reasons for this. First, the main thrust of the new law is aimed at personal injuries not property. Second, business organisations can be expected to insure against damage to their own property and spread the cost of the insurance to their customers. Private consumers may also buy insurance to cover damage to their property, but often do not do so.

Financial limit for property damage

Section 5(4) provides for a financial limit of £275 (excluding interest) and only above this figure is damage to property recoverable. This is obviously intended to exclude small property damage claims. Thus if the damage done to the property of the claimant is worth more than £275, the whole amount is recoverable including the first £275. Presumably in considering the application of this section the net figure after a deduction for contributory negligence is the correct approach.

Burden of proof

Lord Bridge in *Murphy* v. *Brentwood DC* (1991) gave the example of damage to a house, caused by the explosion of a boiler due to the negligent manufacture of the boiler, as recoverable from the manufacturer by the owner of the house in tort on *Donoghue* v. *Stevenson* principles. However, provided the damage exceeds £275 the producer of the boiler would also be liable under the 1987 Act. The advantage of an action under the 1987 Act, provided the boiler was not supplied by virtue of the creation or disposal of an interest in land (see later section 46(4) of the 1987 Act), is that it is not dependent upon proof of negligence by the producer. Subsection (1) of Section 4 (Defences) states that the burden of proof of establishing any of the defences rests on the defendant.

Defences

Defences to the strict liability under the Act are set out in section 4(1).

Compliance with enactments
Section 4(1)(a) makes it a defence 'that the defect is attributable to compliance with any requirement imposed by or under any enactment or with any Community obligation'. It should be noted that it will not be sufficient to show that the product complies with safety regulations under, for example, the Construction Products Regulations 1991. It is necessary to show that the defect arose through compliance with the law and that compliance was mandatory, and that the law could not be met by some other steps which would have rendered the product safe. Furthermore it seems that the relevant regulations must specify the method of manufacture; mere authorisation of a product, e.g. by the British Standards Institution, is not sufficient. If legislation states that additives must be in food and the food is defective because of the additives that would be a defence. In practice the defence seems to have a very limited application.

The person sued did not supply the product
Section 4(1)(b) makes it a defence 'that the person proceeded against did not at any time supply the product to another'.

Section 46(1) deals with the meaning of 'supply', and it reads as follows:

'Subject to the following provisions of this section, references in this Act to supplying goods shall be construed as references to doing any of the following, whether as principal or agent, that is to say –
(a) selling, hiring out or lending the goods;
(b) entering into a hire-purchase agreement to furnish the goods;
(c) the performance of any contract for work and materials to furnish the goods;
(d) providing the goods in exchange for any consideration (including trading stamps) other than money;
(e) providing the good in or in connection with the performance of any statutory function; or
(f) giving the goods as a prize or otherwise making a gift of the goods; and, in relation to gas or water, those references shall be construed as including references to providing the service by which the gas or water is made available for use.'

Subsections 3 and 4 of section 46 read as follows:

'(3) subject to subsection (4) below, the performance of any contract by the erection of any building or structure on any land or by the carrying out of any other building works shall be treated for the purposes of this Act as a supply of goods in so far as, but only in so far as, it involves the provision of any goods to any person by means of their incorporation into the building, structure or works.
(4) Except for the purposes of, and in relation to, notices to warn or any provision made by or under Part III of this Act, references in this Act to supplying goods shall not include references to supplying goods com-

prised in land where the supply is effected by the creation or disposal of an interest in the land.'

The conclusion from these provisions seems to be that the supply by a builder of a building where the supply is effected by the creation or disposal of an interest in the land, does not mean a supply under the Act. This would allow the builder to use the defence under section 4(1)(b), but where the goods supplied are incorporated into the building erected under contract, the Act applies to those goods, e.g. cement.

Section 4(1)(b) limits the scope of liability as it applies if the producers can show that they did not put the product into circulation. Thus all injuries to employees or others caused by manufacture of the product, if occurring on the employer's premises, come within this defence.

Other defences under section 4(1)
The only other defences which are likely to affect the construction industry directly are section 4(1)(d), that the defect did not exist in the product when it was supplied, and section 4(1)(e), for those contractors using very technical methods of production, that the state of the art is such that other producers would not have discovered the defect.

Passing responsibility under section 4(1)(f)
Finally it is possible for a builder to be indirectly affected as section 4(1)(f) provides that the manufacturer of a component is not liable for a defect in the finished (the subsequent) product which is wholly attributable to the subsequent product or compliance with the instructions given by the producer of the subsequent product. For example the supplier of glass may be bound by the instructions given by the builder, yet the thickness ordered may be unsafe for its purpose.

(12) The Construction Products Regulations 1991

These regulations came into force on the 27 December 1991, and implemented an EC Directive on the approximation of laws, regulations and administrative provisions of the Member States relating to construction products. They do not apply to any construction product which was supplied for the first time in the Community before the above date. Furthermore the method of enforcement is by means of the criminal law plus administrative measures, such as the serving of a notice by the Secretary of State prohibiting a person from supplying a product or requiring him to publish a warning about products supplied. Unfortunately there is no civil remedy for breach of the regulations equivalent to section 41 of the Consumer Protection Act 1987, which makes it a breach of statutory duty for any person to contravene the safety regulations made under section 11 of the 1987 Act. Nevertheless the 1991 regulations could be taken into account, for example, in deciding whether there has been a breach of section 14 of the Sales of

Goods Act 1979 (as amended) concerning the satisfactory quality including the fitness for its purpose of a construction product. Presumably also evidence of a conviction under these Regulations could be given in civil proceedings.

(13) The Unfair Terms in Consumer Contracts Regulations 1994

These regulations implemented an EC Directive into English law with effect from 1 July 1995. Until these regulations came into force the Unfair Contract Terms Act 1977 did not pose a problem for the insurance industry as it did not apply to insurance contracts. However, these 1994 Regulations do apply to insurance contracts, but unlike the 1977 Act, which applies the concept of reasonableness (see Appendix 5), it requires 'fairness'.

It seems that insurers should review their contract wording to ensure it is 'fair' in accordance with the Regulations, which also require that contracts are written in plain, intelligible language. Apart from an unfair term being invalid, if it is central to the entire contract the whole insurance policy could be involved. There is also a possibility of a breach of the Regulations giving a right to damages as a result of the unfair term.

However, Regulation 3(2) states that the Court is not permitted to assess the fairness of any term which:

(a) defines the subject matter of the contract; or
(b) concerns the adequacy of the price or remuneration as against the goods or services sold or supplied.

A term will be unfair if it is contrary to the requirement of good faith and causes a significant imbalance in the parties' rights and obligations under the contract to the detriment of the consumer, having regard to all the circumstances at the time the contract was made. The definition of 'consumer' is limited to a natural person who is acting for purposes outside his business. This will exclude most of the higher echelons of the construction industry but does involve those dealing with private individuals.

(14) The Sale and Supply of Goods Act 1994

This Act came into force on 3 January 1995, and the main effect was to replace the term 'merchantable quality' in the Sale of Goods Act 1979 by 'satisfactory quality'. Furthermore the 1994 Act makes it clear that a consumer will not be deemed to have accepted goods because he has stated that he has accepted them (e.g. signing an acceptance note) or has done any act inconsistent with the ownership of the seller unless he has had a reasonable opportunity to examine the goods. This would include ascertaining whether they were of satisfactory quality.

Chapter 2

The main JCT forms and insurance law

2.1 Amendments to the 1980 Edition

The Joint Contracts Tribunal (JCT), which prepares and publishes the main building contract forms most commonly used, had at the time of the issue of the Standard Form of Building Contract (1980 Edition) eleven constituent bodies. The present Joint Contracts Tribunal is generally representative of public sector employers, contractors, sub-contractors and architects. They have to reconcile the various viewpoints of these bodies, and obtaining their agreement takes time; this is reflected in the lengthy period required for the drafting of a new contract or revised edition.

It took nearly two years before the alterations appeared in Amendment 2 issued in November 1986 for incorporation in the January 1987 Revision. These alterations were mainly to the liability and insurance clauses of all versions of the standard forms of the JCT contract. As anticipated in the first edition of this book the basic alteration is for clause 22 of JCT 80 (Insurance of the Works) to call for all risks insurance in place of the fire and special perils cover required under the original 1980 Edition (except in the case of clause 22C.1). This reflects current insurance practice and the opportunity was also taken to remove some inconsistencies in the 1980 provisions.

2.2 The main contract forms

Figure 2.1 shows the main standard forms of JCT or related contracts.

As the purpose of this book is to consider the indemnity (or liability) clauses, the responsibility for the works clauses, and their complementary insurance clauses, the above contracts issued by the JCT and the BEC will be reviewed in the light of this purpose:

The Standard Form of Building Contract 1980 Edition

This contract with quantities, in the private or local authority version (where it differs), is intended for use for major works. This will be referred to as JCT 80, and will include all amendments to date of writing unless otherwise stated.

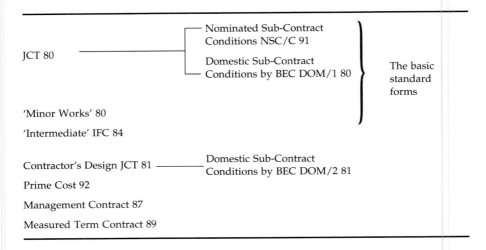

Figure 2.1

Sub-contract forms

The JCT produce a Nominated Sub-Contract form NSC/C and Building Employers' Confederation (BEC) print the sub-contract forms for domestic sub-contractors DOM/1 for use with JCT 80 and DOM/2 for use with JCT 81 (Contractor's Design).

Amendment 10 issued by the JCT in March 1991 replaced the old 'basic' and 'alternative' methods of nomination of sub-contractors, with a single method, which avoids the to-ing and fro-ing of the Tender NSC/1 required by the 'basic' method. This 1991 procedure in no way changes the liabilities of the various parties. The action required by each participant under the 1991 procedure is briefly explained in Chapter 15.

Minor Works and Intermediate Forms

Apparently the Agreement for Minor Building Works was originally suggested by Local Authority Associations represented on the Joint Contracts Tribunal who indicated a need for a standard short form for maintenance work. However, it is also used by private employers. This is a simplified version of the JCT form for minor building works or maintenance work on an agreed lump-sum basis where an architect/supervising officer has been appointed on behalf of the employer and where bills of quantities are not provided. It is the responsibility of the architect to advise his client as to the suitability of the form in any particular case.

This Minor Works Agreement gained considerable popularity with the introduction of JCT 80. In fact because the Minor Works form was being used for larger works for which it was not suitable, the JCT decided to draft an 'Intermediate Form'.

JCT Form of Prime Cost Contract

This contract provides for the remuneration of a contractor by repaying him his costs and, in addition, a fixed fee for his services. The intention is that this form should be used when the exact nature of the work is unknown to a greater or lesser degree at the time the contract is executed.

Management Contract

This contract is part of a package which consists of this contract, the works contract, standard warranties for client, and related supplements, plus guidance notes. The JCT recommends its use for large and complex contracts.

The contractor provides 'management' services instead of full responsibility for the performance of the work. The contractor is paid prime cost (which includes amounts payable under sub-contracts) plus the management fee. The main restriction is that the prime cost is to exclude costs arising from the management contractor's negligence. Where default occurs by a sub-contractor (called a works contractor), the management contractor is only liable if the sub-contractor can also be held liable, but is relieved to some extent by a complex set of relief provisions.

Measured Term Contract

This contract is used for the maintenance of buildings.

Standard Form of Building Contract with Contractor's Design, 1981 Edition

This form is generally suitable for use where the employer has indicated his requirements in writing to a contractor who in response has submitted proposals for the design and tendered for the work. The contract wording generally follows JCT 80 but with many alterations. There is no provision for an architect or quantity surveyor; consequently, many of the architect's functions under JCT 80 are given to the employer under this 'Design and Build' contract.

The contract does not include any requirement that the liability under clause 2.5 (contractor's design warranty) is to be insured, probably because insurance amounting to a total guarantee of the design in all circumstances is not available.

2.3 *General points concerning contract conditions*

Pattern of indemnity, responsibility and insurance clauses

The usual pattern of these clauses in a construction contract is for a liability (indemnity) or a responsibility clause to be followed by an insurance clause concerning that liability or responsibility. 'Liability clause' in this context means a clause imposing a legal liability to indemnify the employer, and 'responsibility clause' means a clause imposing a responsibility for loss or damage to the works. The ICE Conditions are a good example of this pattern but the JCT contract varies the pattern.

The liability clause 20 concerning the contractor's liability to indemnify the employer for personal injury or damage to property, and the complementary insurance clause 21.1, follow the usual pattern, but sub-clause 21.2.1 is only an insurance clause to protect the employer. Clause 22A merely requires the contractor to insure the works. With the removal of the phrase 'at the sole risk of the Employer' from clauses 22B and 22C leaving the responsibility of the employer merely to insure, makes the responsibility otherwise for the works etc. less clear cut. The fact remains that any shortfall in the insurance leaves the party responsible for the insurance also responsible for the shortfall so far as the works etc. are concerned. See also under the heading 'Effect of the Unfair Contract Terms Act 1977' below.

Failure to print in the contract insurance clauses all the common exceptions

Some comparatively successful attempts have been made, after consultation with the British Insurance Association, now the Association of British Insurers, to match the insurance clauses with the usual type of cover granted by insurers, e.g. sub-clause 21.2.1. Even more successful was the introduction of a definition of 'All Risks Insurance' given in clause 22.2 of the 1986 amendment. This assists those who are responsible for checking the contractor's contract works insurance. Unfortunately this aspect was not extended into the insurance requirements concerning liability insurance in clause 21. On this matter see the writer's PhD thesis, *An improved method of requesting insurance under UK construction contracts*. (A copy of this work is available at the City University library in London.) Since the 1986 amendment it has been made clear that liability insurance cover for personal injury from nuclear risks is not required as it is unobtainable. Similarly with sonic waves risks, as they also are part of the 'excepted risks' defined in clause 1.3. JCT 80 has always excluded loss or damage to the works or any other property so far as nuclear and sonic waves risks are concerned.

Lack of precision in drafting

It is surprising, after so many years of standard forms of contract, that words having a legal meaning are used in conjunction with general words having no legal meaning, thus leaving the reader in doubt as to whether the general words add anything or are superfluous. In the JCT contract there are two such phrases used in the liability clause 20 and the complementary clause 21. The phrases are 'act or neglect' in clause 20.1, and 'negligence, breach of statutory duty, omission or default' in clause 20.2. Both of these sub-clause numbers are referred to in clause 21.1, and the clause number 20.2 also appears in sub-clause 21.1.1.1. Now 'negligence' or 'neglect' and breach of statutory duty have a legal definition but 'act', 'omission' and 'default' do not. If it is intended to include other torts they should be specifically stated, and the words 'act', 'omission' and 'default' should not be included. In fact the exception 'unless due to any act or neglect of the Employer...' is apparently limited in clause 20.1 to common law negligence by virtue of *Hosking* v. *De Havilland Ltd* (1949) and *Murfin* v. *United Steel Companies Ltd* (1957), and details of the former case will be given when discussing clause 20 in detail (see Chapter 3). However, reference to the courts should not be necessary if the clauses were drafted properly in the first place.

Effect of the Unfair Contract Terms Act 1977

A short summary of the purpose of each section of Part I of this Act is given in Appendix 5. Usually negotiated standard forms of contract like the JCT Form do not avoid liability in respect of the negligence of either party. However, there are exceptions, and two of the basic sections of this Act must be understood. This Act makes such exclusion terms concerning negligence resulting in death or personal injury invalid, or if other loss or damage results, subject to a test of reasonableness concerning that term. Thus this Act could by section 2 abrogate that term. Furthermore, section 3 requires a test of reasonableness in deciding whether a term avoiding liability for breach of contract, in written standard terms of business, is valid.

It is unlikely that the decision in *Archdale & Co Ltd* v. *Comservices Ltd* (1954) will be an acceptable interpretation of clause 20.2 in conjunction with clauses 22B or 22C since the 1986 amendment. The point is that the words 'shall be at the sole risk of the Employer as regards loss or damage by fire (later Clause 22 Perils)' no longer appear. This was the main reason why the employer was held responsible for negligent fire damage to the works by the contractor. However, it would not seem to be reasonable to hold the employer responsible for such damage under the 1977 Act without the above wording (see also Chapter 13).

'Clause 22 Perils' has now been deleted from the contract under the 1986 amendment and, apart from clause 22C.1, 'All Risks Insurance' replaces the requirement of a fire and special perils policy in clause 22. In the House of Lords case of *Scottish Special Housing Association* v. *Wimpey* (1986), which was heard before the 1986 amendment was involved, the same decision was reached as the circumstances were similar to those in the *Archdale* case. Thus Wimpey was allowed to avoid liability for their own negligence. It is also worth noting that the House did not consider the Unfair Contract Terms Act with its reasonableness test. The House merely considered the ordinary meaning of the words in their contractual setting, so the words 'at the sole risk of the Employer' were vital. In spite of the above mentioned amendments, because of other alterations in 1986 to the JCT contract, e.g. that 'property' in clause 20.2 does not include the works, it does not follow that the contractor or the sub-contractor will be liable for their negligence in causing loss or damage to the works when clauses 22B or 22C apply (see later the case of *Ossory Road (Skelmersdale) Ltd* v. *Balfour Beatty Building Ltd and Others* (1993).

A rule of construction

The House of Lords in *Smith* v. *South Wales Switchgear (1978)* adopted the following validity test of an indemnity clause from previous cases:

(a) if it contains language expressly exempting a party from his negligence it must apply;
(b) otherwise, consider whether the wording is wide enough to cover negligence – if in doubt, it does not apply to negligence;
(c) if the answer to (b) is 'yes', consider whether damages may be based on grounds other than negligence – if so, it cannot apply to negligence.

Thus, although this decision is sometimes said to be that to contract out of negligence requires the use of the word 'negligence' or a synonym for it (i.e. (a) above), the other principles must not be forgotten, e.g. in *The Raphael* (1982) the words 'any act or omission' could only apply to negligence, (i.e. (c) above).

However, this rule, like the Unfair Contract Terms Act, is not always considered.

2.4 *Insurance law*

The insurance contract has a set of legal rules and sometimes these replace those normally applying to the law of contract.

Utmost good faith

Good faith must be observed in *all* legal contracts; that is to say, the parties must not act fraudulently. However, this does not mean that they are bound to reveal all they know, or ought to know, about the transaction. Contracts of insurance are different because one party to the contract alone (the proposer) knows, or ought to know, all about the risk proposed for insurance, and the other party (the insurer) has to rely mainly upon the information given by the proposer in his assessment of that risk. For this reason, contracts of insurance are contracts *uberrimae fidei*, of the utmost good faith.

Disclosure

This means that all material facts must be disclosed. Section 18(2) of the Marine Insurance Act 1906 states that:

> Every circumstance is material which would influence the judgment of a prudent insurer in fixing the premium or determining whether he will take the risk.

In this respect the law in non-marine insurance is the same as that in marine insurance. Because the statement is ambiguous in that it could mean something which 'influences' the insurer on a 'required to know' basis but later discarded as unimportant, or something which is a deciding factor in determining the conclusion of the insurance contract or its terms, the lawyers and the courts have had a field day over the last decade.

The Court of Appeal in *CTI* v. *Oceanus* (1984) took the broad view of the quotation, i.e. taken into account but later discarded. Some felt that this imposed an obligation on the insured to disclose virtually endless material about the insured's past.

The House of Lords considered the matter in *Pan Atlantic Insurance Co Ltd* v. *Pine Top Insurance Co Ltd* (1994) and established:

(1) An insurer can avoid liability only if two requirements are satisfied, namely
    (a) the judgment of a *prudent* insurer would have been influenced by the information not disclosed, and
    (b) the insurer *in question* was in fact induced by the insured's presentation of the risk, to enter into the insurance contract.
(2) Three of the judges held the test for the phrase 'would influence the judgment of a prudent insurer' in 1(a) above is whether that judgment would have been affected by the information concerned. Lord Goff described it as 'an effect on the mind of the insurer in weighing up the risk'. The minority held the test was stricter and required proof that if a prudent insurer had known of the undisclosed fact he would either have declined the risk or required a higher premium.
(3) In 1(b) above it is necessary for the insurer to show he had actually been induced to enter into the contract by the insured's withholding or misstating information. What was required was inducement and proof that, but for the inducement, the contract would not have been made.

In conclusion it seems that while the majority of the House sided with the *CTI* decision in 1(a) above, the House restricted the *CTI* decision in 1(b) by introducing the novel feature of 'inducement' of the particular insurer to enter into the contract. In *St Paul Fire and Marine Insurance Co (UK)* v. *McConnell Dowell Constructors* (1993) the plaintiff provided construction works for the defendant. The risks were presented in error as piled foundations were stated to be used, but spread foundations were in fact employed as a result of an architect's report. There was subsidence and the plaintiff purported to avoid the policy. It was held in favour of the plaintiff that a prudent insurer would view the undisclosed information as increasing the risk. While this case was heard before the House of Lords decision in the *Pan Atlantic* case there seems no doubt that the same decision would have been made if this Court of Appeal case had been heard after the House of Lords decision.

The proposal form

Consequently, this duty is not necessarily discharged by the proposer who replies truthfully to all the questions on the proposal. Any other material facts which might affect the insurer's mind in considering the risk and his inducement to enter into the contract must be disclosed. This means any material facts not covered by questions on the proposal (see *Hair* v. *Prudential Assurance Company Ltd* (1982)).

The following list, which is not exhaustive, should assist in the completion of the proposal form:

- Answer all questions fully and completely honestly. Although this may seem elementary, it is often not complied with. Even if questions are not applicable, this should be stated, explaining why the question does not apply unless it is obvious.
- Disclose all matters which a prudent insurer would wish to consider in deciding whether to offer cover. If in doubt disclose the matter.
- Obtain a specimen of the policy required. Check that the exclusions and conditions are understood and are acceptable.
- Any special explanations required concerning the cover should be obtained in writing and the correspondence retained, similarly in the case of policy extensions.
- Sums insured and limits of indemnity must be adequate.
- Keep a copy of the completed proposal.

Continued monitoring

Throughout the policy period the insured should:

- Notify any change in risk which is basic to the cover provided, e.g. an extension of geographical, or the type of, activity.
- Notify claims immediately. If in doubt, notify the circumstances concerning a possible claim.
- In the case of liability insurance claims, make no admission, offer, promise or payment to the third party and pass relevant correspondence to the insurer immediately.
- On renewal the proposal answers should be checked and all the changes or requirements notified.

If the above points are followed, repudiation of policy liability by the insurer on the grounds of non-disclosure is unlikely.

Other principles

Insurable interest

The second principle of insurance is that the insured must have an insurable interest in the subject matter of the insurance, whether it is property, life or limb, or potential liability devolving upon the insured. That is to say, he must bear some relationship, recognised by law, to the subject matter whereby he benefits by the safety of the property, life or limb or freedom from liability, and be prejudiced by any loss, damage, injury or creation of liability.

Indemnity

Thirdly, all the policies mentioned in this book are policies of indemnity, which means that the object of these insurance contracts is to place the insured as nearly as possible in the same financial position after a loss as that occupied immediately before the happening of the insured event. It would be against public policy to allow the insured to make a profit out of the occurrence.

Subrogation

Fourthly, there is the principle of subrogation. This is the right which the insurer has of standing in the place of the insured and availing himself of all the rights and remedies of the insured, whether already enforced or not, but only up to the amount of the insurer's payment to the insured. This right of subrogation is exercisable at common law after the insurer has paid the claim. However, a condition of the policy may entitle the insurer to exercise the right before the payment is made.

In the construction industry it is sometimes the practice to insure many of the parties involved in the construction project as joint insureds under the one policy. In this event the principle of subrogation cannot operate to allow the insurer, after paying one insured, to recover from a joint insured (see *Petrofina (UK) Ltd* v. *Magnaload Ltd* (1983)).

Contribution

Finally, the principle of contribution, like subrogation, is a corollary of indemnity. It only applies between insurers. Contribution is the right of an insurer who has paid under a policy to call upon other insurers equally or otherwise liable for the same loss to contribute to the payment. Before the principle can be applied, the insurances called into contribution must cover (a) the same interest, (b) the same subject matter, and (c) the same peril. Contribution will not operate if one policy clearly states that it only applies after any other double insurances have been exhausted, and that other insurance does not contain the same clause. In practice, policies usually contain a non-contribution clause of some kind, but some are more strictly worded than others (see *State Fire* v. *Liverpool London & Globe Insurance Co Ltd* (1952).

The statements made above have been verified in the Scottish appeal case of *Steelclad Ltd* v. *Iron Trades Mutual Insurance Co Ltd* (1984). However, the interesting part of that case was the court's view of the words in the contribution condition of the policy of the defenders which refers to loss or damage insured by any other policy 'effected by the insured . . . on his behalf' when the defenders would not be liable except in respect of any excess beyond the amount payable under such other policy. The pursuers were the insured under the Iron Trades policy and were sub-contractors on a project where the employer having the work done had arranged a project policy covering the employer, all contractors and all sub-contractors. The contribution clause in that policy was worded exactly the same as the Iron Trades policy except for the phrase quoted above. The court considered the project policy was not a policy effected by the insured on his behalf. Because the phrase was ambiguous, it was construed *contra proferentem* (see below) against the insurer who had refused to contribute. The project policy insurers had already paid almost half of the claim. The court did not have to go any further, having come to a decision on this wording. However, they did in effect say that, even without this phrase or if the project policy had been effected on the pursuers' behalf, they *would still not allow the two policy conditions to cancel the cover given* (by ignoring in each case the condition which was worded exactly the same way in the other policy), with the result that if the loss is covered elsewhere, it is covered nowhere.

The correct decision, and surely the intention, is that each policy should contribute subject always to a rateable proportion condition. Basically, an insured must get an indemnity whether he insures once or twice the same subject matter against the same peril, subject always to the other terms and conditions of the policy or policies. It is rather strange that the matter should ever have been in doubt.

2.5 *The insurance contract*

An insurance contract is an agreement whereby one party, the insurer, in return for a consideration, the premium, undertakes to pay the other party, the insured, a sum of money or its equivalent upon the happening of a specified event which is against the insured's financial interest.

Component parts of a policy

The contract is normally contained in a policy. This document is evidence of the contract which has usually come into existence before the policy is issued. Policies vary with different classes of insurance, and those dealt with in this book can be divided into the following sections:

● Recital clause, which refers to a schedule for details (see below). This clause also refers to the proposal form and its declaration as being the basis of the contract. However one qualifies 'the event', it must be uncertain.

- Operative clause, which describes the cover which is the subject matter of the insurance.
- Exceptions which help to describe the cover by stating what is excluded.
- Signature clause on behalf of the insurers.
- Schedule, which contains the names of the parties, the address of the insured, the period of insurance, possibly the business of the insured, the sums insured or limit(s) of indemnity (in liability insurance), and the premium payable.
- Conditions, which limit the insured's legal rights, stipulate the various things the insured may or may not do and sometimes express or amend the common law or indicate an agreed state of affairs.
- Endorsements, which vary the standard cover so that it is tailor-made to fit the insured's work.

Chapter 5 explains the classes of policies, and Chapters 6 and 7 deal with specific policies considered in this book. The policy may be amended from time to time, with the agreement of the parties, by endorsements which are attached to the policy.

Conditions

It is usual to classify conditions (whether expressed or implied) as:

(a) Conditions precedent to the policy, e.g. all material facts must be disclosed during negotiations preceding the insurance contract.
(b) Conditions subsequent of the policy, e.g. notice by the insured of a change of risk during the period of the policy.
(c) Conditions precedent to liability, e.g. the notice of loss by the insured.

The following is the result of this classification:

(a) Conditions precedent to the policy must be observed for the insurance to be valid from the beginning.
(b) Conditions subsequent of the policy refer to matters arising after the contract has been completed and affect the validity of the policy from the date of breach of the condition.
(c) Conditions precedent to liability only affect the claim which the breach of the condition concerns, the policy remaining in force.

A condition by which the insured undertakes that some particular things will be done or not done or that a state of affairs exists or shall continue to exist is referred to as a warranty or a continuing warranty. Breach of such condition will probably allow the insurer to repudiate liability, as a warranty must be complied with strictly and literally.

Occasionally a so-called condition in the policy has been held to be a mere stipulation. Thus it has been stated that the condition in the employers' and public liability policies requiring the insured to keep a proper wages book in order to return his annual wage roll on which the premium is based is a mere stipulation, a breach of which results in the insurer possibly obtaining damages as compensation but not allowing him to repudiate policy liability.

There is a rule of evidence among many concerning the interpretation of the insurance policy, which has already been mentioned in this chapter as it occasionally arises. It is called the *contra proferentem* rule and states that in the event of ambiguity a document will be construed strictly against the party who has drawn it up. Thus the interpretation less favourable to the insurer will be taken.

From what has been said under the last two headings concerning non-disclosure and breach of warranty, it will be appreciated that insurance law is tilted in favour of the insurer. However, the Law Commission's Working Paper No 73, *Insurance Law, Non-disclosure and Breach of Warranty* recommended reforms to improve the insured's position. This was followed by a Final Report on the same subject and on the same lines, with a draft Insurance Law Reform Bill annexed. However, this was in 1979 and 1980 and no steps have been taken to progress the recommendations.

Chapter 3

Injury to persons and property: clause 20

20.1 The Contractor shall be liable for, and shall indemnify the Employer against, any expense, liability, loss, claim or proceedings whatsoever arising under any statute or at common law in respect of personal injury to or the death of any person whomsoever arising out of or in the course of or caused by the carrying out of the Works, except to the extent that the same is due to any act or neglect of the Employer or of any person for whom the Employer is responsible including the persons employed or otherwise engaged by the Employer to whom clause 29 refers.

20.2 The Contractor shall, subject to clause 20.3 and, where applicable, clause 22C.1, be liable for, and shall indemnify the Employer against, any expense, liability, loss, claim or proceedings in respect of any injury or damage whatsoever to any property real or personal in so far as such injury or damage arises out of or in the course of or by reason of the carrying out of the Works, and to the extent that the same is due to any negligence, breach of statutory duty, omission or default of the Contractor, his servants or agents or of any person employed or engaged upon or in connection with the Works or any part thereof, his servants or agents or of any other person who may properly be on the site upon or in connection with the Works or any part thereof, his servants or agents, other than the Employer or any person employed, engaged or authorised by him or by any local authority or statutory undertaker executing work solely in pursuance of its statutory rights or obligations.

JCT 80 clause 20 is not an insurance clause as it does not require the contractor to effect any insurance cover. It deals with the liability of the contractor to indemnify the employer. Incidentally, although, strictly speaking, an indemnity only applies after judgment has been found against the employer, he can take third party proceedings against the contractor before such judgment (see *County & District Properties* v. *Jenner & Sons Ltd* (1974)).

It is intended to analyse the phrases in turn, taking first sub-clause 20.1.

3.1 Injury to persons (clause 20.1)

The liability placed upon the contractor and the indemnity he gives to the employer are wide. The five words emphasised in the following extract indicate the extent of the liability and indemnity.

> ... *any* expense, liability, loss, claim or proceedings *whatsoever* arising under *any* statute or at common law in respect of personal injury to or the death of *any* person *whomsoever*

The opening phrase would include any consequential loss.

The words 'under any statute' call for a reference to Chapter 1 under the heading 'Breach of statutory duty'. Probably the most important of the statutes mentioned in that chapter are the Fatal Accidents Act 1976 and the Occupiers' Liability Act 1957. The words 'any person whomsoever' include the employer himself, his servants and agents as well as the more obvious potential claimants.

In the proviso

> ... except to the extent that the same is due to any act or neglect of the Employer or of any person for whom the Employer is responsible including the persons employed or otherwise engaged by the Employer to whom clause 29 refers.

The phrase 'except to the extent that the same is' in place of the word 'unless' is apparently intended to make clear that the indemnity to the employer is still partially to apply, even if some part of the personal injury or death is due to the act or neglect of the employer.

In Chapter 2 under the heading 'Lack of precision in drafting', a criticism of the phase 'act or neglect' was made although the cases quoted there decided that the phrase was limited to common law negligence, and is not applicable to a breach of statutory duty. Thus the contractor would still have to indemnify the employer where the claim by the injured person is based on a breach of statutory duty by the employer. *Hosking* v. *de Havilland Ltd* (1949), which dealt with sub-clause 14(a) of the 1939 RIBA contract (containing a similar wording to sub-clause 20.1 of JCT 80), decided that the words 'act or neglect' do not include a breach of statutory duty, as there would be very little for the indemnity (to the employer) to apply to if these words in the exception included such a breach by the employer as well as his common law negligence.

The Court of Appeal in *Murfin* v. *United Steel Companies Ltd* (1957), when interpreting a similar clause, drew a distinction between statutory liability and negligence and approved of the decision in *Hosking*. In the latter case the plaintiff (an employee of the defendants) was injured when a plank placed over a duct by a contractor broke. The plaintiff stepped on the plank to cross the duct. The court held the contractor liable at common law and the defendants in breach of the Factories Act current at the time. However, the contractor was held liable for the total damages and costs because of the contract conditions, otherwise the damages would have been apportioned between the contractor and the defendants.

If it can be presumed from these cases that the word 'act' is superfluous, then an act of nuisance or trespass to the person by the employer causing injury would result in his obtaining an indemnity under this clause.

This indemnity not only excludes injuries or death 'due to any act or neglect of the Employer' but also 'of any person for whom the Employer is responsible'. This means not only his employees, for whom he is vicariously responsible, but also in certain circumstances his agents. The JCT form in clause 29.3 states:

> Every person employed or otherwise engaged by the Employer as referred to in clauses 29.1 and 29.2 shall for the purpose of clause 20 be deemed to be a person for whom the Employer is responsible and not to be a sub-contractor.

Clauses 29.1 and 29.2 refer to work not forming part of the contract either provided or not provided for in the contract bills, although in the latter event the contractor's consent (not to be unreasonably withheld) is required.

The obvious question is whether the architect and quantity surveyor are persons for whom the employer is responsible. It appears that they are in certain circumstances, for example in the functions they perform as agents of the employer, either implied or expressed under the contract. Thus the employer will not be able to deny the architect's or quantity surveyor's authority in this respect. However, in *Clayton* v. *Woodman & Son (Builders) Ltd* (1962) (where an employee of the builder was injured), an argument that an employer was liable for the negligence of an architect employed by him in the normal way to prepare designs and specifications, and to supervise certain building work, was rejected in positive terms by Salmon J, who said:

> 'The plaintiff contends that the second defendants [the employers] are vicariously liable for the negligence of their architects, the third defendants. In my judgment this contention is scarcely arguable. It is true that for some purposes an architect is the agent of the building-owner, e.g. for the purpose of ordering materials: *Wallis* v. *Robinson* (1862), 3F. & F. 307. Nevertheless an architect is clearly an independent contractor. The building-owner has no control over the manner in which the architect does his work.'

Atiyah in his book *Vicarious Liability in the Law of Torts* criticises the use of the control test in this statement but otherwise agrees with this decision.

On appeal this question of the employer's vicarious liability did not arise but reference was made to the architect's position as an agent to make sure that the employer will have a building properly constructed in accordance with the plans, specification, drawings and any supplementary instructions issued by the architect.

Therefore on balance it is difficult to visualise the employer being responsible for injury to a third party negligently caused by the architect, but if, as illustrated by Salmon J in the quotation above, an expense is incurred by a third party owing to negligence of the architect in ordering materials as agent of the employer, then the latter would be liable.

3.2 Damage to property (clause 20.2)

It seems strange to the layman (including members of the insurance industry) to use the word 'injury' when referring to property, which the heading to this sub-clause does, hence the use of the word 'damage' in its place. 'Injury' denotes personal injury and can cause confusion if used where it is intended to refer to property only. However, it has to be admitted that sometimes injury to property is spoken of when referring to the impairment of its value by nuisance, e.g. noise and smells.

This clause does not include the works in the phrase 'property real or personal' and, probably because it was never intended to do so, this aspect is now made clear in sub-clause 20.3 (see below).

Consequently the first line of sub-clause 20.2 reads: 'The Contractor shall, subject to clause 20.3 and, where applicable, clause 22C.1, be liable for...' The reason for the inclusion of clause 22.C.1 is given later. This sub-clause then continues on the same lines as in the previous edition except that the same phrase is used as in 20.1, namely 'and to the extent that the same is...', which again is used with the intention of making it clear that the contractor is liable, and the indemnity to the employer is still in force, even if only a part of the injury or damage is due to the negligence of the contractor. Whether it is legally interpreted that way remains to be seen.

Contractor's wrongdoing

The important question under this heading is what is the meaning of 'negligence, breach of statutory duty, omission or default of the Contractor'. There is no difficulty about negligence and breach of statutory duty as they are defined in Chapter 1. From the facts of a particular situation it is usually clear whether a duty of care is owed and if so whether it has been breached. Similarly for a breach of statutory duty. The remaining concepts are more difficult to explain.

'Omission' is included in the definition of negligence and in that respect there is an overlap with the first term. On the other hand, there can be an omission in deciding whether there is default (see the definition later) so 'omission' does seem to be superfluous to the whole phrase. In other words it is difficult to show that the word 'omission' adds anything to 'negligence' and 'default'.

In *City of Manchester* v. *Fram Gerrard Ltd* (1974) Kerr J quoted from the decision of Parker J in *Re Bayley-Worthington and Cohen's Contract* (1909) where the meaning of 'default' was considered. Kerr J adapted Parker J's words to the phrase under discussion. This then came from clause 14(b) of the 1961 Standard Form of Building Contract, which was very similar to clause 20.2 of JCT 80. Kerr J said:

'I think that default would be established if one of the persons covered by the clause either did not do what he ought to have done, or did what he ought not to have done, in all the circumstances, provided of course that the remaining provisions of the clause are also satisfied and that the conduct in question involves something in the nature of a breach of duty so as to be properly describable as a default.'

This is a very wide definition and apart from including a breach of a duty to take care (already covered by the word 'negligence'), includes a breach of contract and some breaches of statutory duty. However, this is where any attempt at exploration becomes confusing because some breaches of statutory duty are irrespective of fault, which is why the word 'some' was used in the previous sentence.

'Fault' according to the *Concise Oxford Dictionary* is a synonym for default. Thus it is doubtful whether 'default' applies to nuisance or trespass to land which apply irrespective of fault. On the other hand, fault is a requirement in trespass to goods. See *National Coal Board* v. *Evans* (1951). Also default could include trespass to the person as default is wide enough to include 'breach of duty', and in *Letang* v. *Cooper* (1965) (see Chapter 1 under the heading 'Trespass') Lord Denning said:

> 'Our whole law of tort today proceeds on the footing that there is a duty, owed by every man not to injure his neighbour in a way forbidden by law.'

Does the word include breaches of duties beyond the law of contract and tort? It is not applicable in the realm of strict liability where fault is not necessary, which is within the area of tort, nor to some breaches of statutory duty; but what about moral duties? Without further case law it is impossible to answer this question but the width of meaning in the word 'default' can clearly be seen as considerable. If only drafters of contracts would adhere to legal terms as pleaded in Chapter 2, then one would be much more certain as to the position.

'Servants or agents'

This phrase 'his servants or agents or of any person employed or engaged upon or in connection with the works or any part thereof, his servants or agents' increases the contractor's common law liability as he is not usually responsible for the torts or default of his sub-contractors, although he is always responsible vicariously for acts committed by a servant acting in the course of his employment. A person employed or engaged upon or in connection with the works may be either a servant or an independent contractor. Such a person may also, at the same time, be an agent. It seems that a separate category of agents if they are not servants or independent contractors has little relevance in the present context. An agent may or may not be subject to the degree of control which will make him a servant. The law of tort here is concerned only to know in any particular case whether he is a 'servant'. If the person concerned is a servant then the only question is whether he was acting within the scope of his employment. If so the contractor is liable for his acts. If he is an independent contractor, i.e. a sub-contractor, the question in tort is as follows.

Early case law established circumstances in the common law of tort where a principal is liable for the defaults of his contractor, which would equally apply to a main contractor being responsible for his sub-contractor, but these do not need to be considered in the context of works done under JCT 80 as this exhaustive phrase makes the main contractor liable, so far as indemnifying the employer is concerned. However, as the cases cast light on this sub-contract relationship they are set out for reference in Appendix 11.

Other persons on site

The contractor's liability extends to the negligence, etc. of 'any other person who may properly be on the site upon or in connection with the Works or any part thereof, his servants or agents'.

As well as the main contractor, the sub-contractor and sub-sub-contractors, accepting an architect, his servants and agents as independent contractors or agents of the employer (see the discussion earlier), there are other people who may properly be on the site in connection with the works, e.g. insurance officials. If these people cause injury or damage to the employer, the contractor must indemnify him. Unauthorised persons such as trespassers would not be 'properly' on the site. However, the contractor's liability is restricted by a final statement:

> ... other then the Employer or any person employed, engaged or authorised by him or by any local authority or statutory undertaker executing work solely in pursuance of its statutory rights or obligations.

This restricts the indemnity given to the employer by the contractor, as it removes the persons mentioned from the indemnity. Thus as well as the employer and his employees, there will be a variety of people engaged or authorised by the employer and by any local authority or statutory undertaker executing work solely in pursuance of its statutory rights or obligations and the negligence, etc. of all these people is not within the indemnity given to the employer. This includes all those persons for whom the Employer is made responsible under clause 29 mentioned in clause 20.1.

Exclusion of the works and site materials from damage to property

'20.3.1 Subject to clause 20.3.2 the reference in clause 20.2 to 'property real or personal' does not include the Works, work executed and/or Site Materials up to and including the date of issue of the certificate of Practical Completion or up to and including the date of determination of the employment of the Contractor (whether or not the validity of that determination is disputed) under clause 27 or clause 28 or clause 28A or, where clause 22C applies, under clause 27 or clause 28 or clause 28A or clause 22C.4.3, whichever is the earlier.

20.3.2 If clause 18 has been operated then, in respect of the relevant part, and as from the relevant date such relevant part shall not be regarded as 'the Works' or 'work executed' for the purpose of clause 20.3.1.

Sub-clause 20.3, as already mentioned, makes it clear that the phrase 'property real or personal' in sub-clause 20.2 'does not include the Works, work executed and/or Site Materials'. Clause 27 deals with determination by the employer and clause 28 with determination by the contractor. Clause 22C.4.3 deals with determination by either party.

The second part of sub-clause 20.3 (20.3.2) states that if clause 18 (dealing with partial possession by the employer) operates then, in respect of the 'relevant part' and as from the 'relevant date', such relevant part shall not be regarded as 'the Works' or 'work executed' for the purpose of the first part of sub-clause 20.3 (20.3.1). This is because such relevant part from the relevant date is treated as if it were third party property and comes within sub-clause 20.2, subject to its other terms applying (see below).

Incidentally, it should be noted that, where clause 22C applies, as clause 22C.1 deals with the employer's obligation under clause 18 to take out a joint names policy after the relevant date, for loss or damage by the specified perils to the 'existing structures' (as distinct from works in or extensions to existing structures where all risks cover is required), clause 22C.1 will also apply to the relevant part: i.e. the relevant part, which was works, becomes existing structures (see sub-clause 18.1.3).

On the other hand there is no requirement under clause 22A or 22B (new buildings) for the employer to insure the 'relevant part' after the 'relevant date' if clause 18 operates, as there is where clause 22C applies.

3.3 General matters concerning clause 20

The essential difference between sub-clauses 20.1 and 20.2

It may be wondered why the onus of proving the application of the exception in sub-clause 20.1 is on the contractor, whereas in sub-clause 20.2 the onus of proving that the final proviso operates is on the employer.

It is simply a matter of the way the sub-clauses are worded. A similar situation concerning onus of proof exists under insurance policies, where the onus is upon the insured to prove that the circumstances of an incident come within the wording of the operative clause, and the onus of proving that an exclusion applies rests on the insurers.

It has been suggested that the reason for the difference in the two sub-clauses is the desire in the public interest to avoid as far as possible disputes over liability for personal injuries, hence the wider responsibility of the contractor; but as considerable damage to adjacent property can be caused as much by faulty design as by the contractor's negligent carrying out of the works, the responsibility in the case of the property damage must be left where it normally lies at law. However, it is emphasised that these points are only suggestions, and the reason for the difference is not known.

A more practical reason may be that some types of work, however carefully executed, involve a risk of damage to adjacent property for which it would be unreasonable for the contractor to pay. On the other hand, claims for personal injuries usually arise from work over which the contractor has control.

Value of clause 20 as an indemnity clause

Liability to a third party

In both sub-clause 20.1 and 20.2, if the contractor is sued in circumstances where the employer is not entitled to an indemnity, i.e. where the injury is due to the negligence of the employer in clause 20.1, or in sub-clause 20.2 where the damage does not arise directly from the contractor's default but from the employer's instructions, recovery from the employer is possible. The contractor in these circumstances could have a right of contribution under the Civil Liability (Contribution) Act 1978 against the employer, which could amount to a complete indemnity.

The use of the words 'to the extent that the same is due to' in both clause 20.1 and clause 20.2 clearly indicates the intention to apply an apportionment between the parties when the indemnity applies, so that the employer and the contractor share responsibility to a third party in either equal or unequal proportions. In the first two of the following cases, which indicate some of the problems that can arise under clause 20, it must be appreciated that there was some doubt as to whether the phrase 'property real or personal' in clause 20.2, or its equivalent in earlier contracts, included the works. Currently clause 20.3 shows it does not.

Liability for fire as between employer and contractor

In *Archdale* v. *Comservices* (1954) the contract was the JCT 1952 Edition and the contractor negligently destroyed the works by fire. The contract provided that the contractor should indemnify the employer in similar fashion to the current clause 20.2, but not the same wording. However, it did provide that if this damage was due to the negligence of the contractor he was liable to pay for that damage. The contract also provided in similar fashion to the current clause 22B, but not the same wording that the works should be 'at the sole risk of the Employer' as regards loss or damage by fire, and that the employer should maintain a policy of insurance against that risk. The court held that the combined effect of these two clauses was for the employer to bear the risk of damage by fire in spite of the contractor's negligence. In *Scottish Special Housing Association* v. *Wimpey* (1986) the same situation arose as in *Archdale* with the same two clauses involved, but this was the JCT 1963 Edition 1977 Revision. The same decision was reached thus allowing Wimpey to avoid liability for their own negligence. The interesting point about the last case is that the House of Lords did not consider either the rule of construction they applied in *Smith* v. *South Wales Switchgear Ltd* (1978) (see Chapter 2), nor the Unfair Contract Terms Act 1977 with its test of reasonableness. They just considered the ordinary meaning of the words, as Steyn LJ said in *E.E. Caledonia Ltd* v. *Orbit Valve Co Europe* (1994) (which was not a JCT case nor did it involve damage to property) 'an aid to the process of construction' does not oust 'the dominant factor', the 'ordinary meaning of the words in their contractual setting'.

A more recent case on similar facts is *Ossory Road (Skelmersdale) Ltd* v. *Balfour Beatty Building Ltd and Others* (1993). Under a JCT 80 contract incorporating the 1986 amendment (which is the current wording), Balfour agreed to construct a three-level retail and car park development and to refurbish the public areas on the concourse shopping centre in Skelmersdale for Ossory as employer. Balfour engaged Briggs Amasco to carry out roofing work. By this sub-contract Briggs were deemed to have notice of the conditions of the main contract and undertook to comply with them so far as they applied to the roofing work. Fire damage occurred to the contract works but there was no allegation of damage to the existing structures although the fire damage would affect the use of the existing structures. The fire engulfed the roof of the newly constructed building.

In order to follow the judge's findings, assuming negligence on the part of Balfour and/or Briggs Amasco, it is necessary to explain some of the terms added to the JCT 1980 contract in the 1986 amendment. Clause 20.3.1 makes clear that 'property' does not include the works, meaning 'the words executed'. Also under clauses 22C.1 and 22C.2 the employer must take out joint names insurance of the existing structures and contents against specified perils and must take out joint names insurance of the works on an all risks basis. Therefore, as the indemnity in clause 20.2 was modified by clause 20.3 (excepting the works) and 'where applicable', clause 22C1 (concerning the insurance of existing structures), the judge decided that Balfour would **not** be liable either for works executed (which was the case in question) or for the cost of any damage to existing structures (which was not the case here) as a result of fire.

The final issue on this case was whether Briggs Amasco owed Ossory a duty of care. The judge followed the decision in *Norwich City Council* v. *Harvey* (1989). JCT 63 applied in that case. It was held that notwithstanding that a sub-contractor is not party to the main contract he may, by application of the contract terms, be relieved of a duty he would otherwise be under. This aspect will be discussed further when considering sub-contractors (see Chapter 15).

In coming to his decision between Ossory and Balfour the judge said:

'In determining whether or not the effect of Clause 20.3 is to exclude any liability and/or indemnity obligations on the part of Balfour to Ossory for damage by fire to the Works or works executed, the provisions of clause 22C.2 can have only limited interpretation relevance since they are not expressly referred to in clause 20.2. I have reached the conclusion that by excluding by clause 20.3 the Works and work executed from clause 20.2 Balfour were under no liability to Ossory in respect of damage or injury to the property and were not required to indemnify Ossory against expense, etc.'

It might be thought that, as the words 'at the sole risk of the Employer' had been removed from clause 22C in the Ossory contract, the result would have been different from the first two cases mentioned above. However, it seems that the removal of the works from the meaning of 'property' in clause 20.2 by 20.3 has resulted in the contractor still being absolved from his negligence in causing damage to the works. In *Scottish Special Housing Association* Lord Keith concluded that the key question which clause 20(C) (now clause 22C) sought to answer was – who should insure against the contractor's negligence? Clearly in that case it was the employer. Possibly this was always the intention but neither clause 22C.1 nor clause 22C.2 say this, they merely require the employer to take out and maintain insurance. However, it seems that responsibility for joint names insurance has the same effect as the words 'at the sole risk of the Employer' did in the earlier cases, where joint names insurance was not required.

Bearing in mind that clauses 22B and 22C require the employer to insure in the joint names of the employer and contractor it might be thought that there is no difference between insuring the works and insuring the contractor's negligence causing the loss or damage to the works, etc. This would overlook the fact that insurance of the works has its limitations and exceptions (see Chapter 7). For example in *Ossory* it was stated that although there was no damage to the existing structures the use of these structures would be affected. Now if a claim for loss of use were made it would not be covered by the contractor's all risks insurance required by clause 22.2 as it would be a consequential loss, not *'physical* loss or damage to work executed' (see Chapter 7). Therefore there is a difference and it is relevant to consider the position of the contractor negligently causing loss or damage to the works, etc.

Altering the printed conditions

The remarks under this heading are not limited to clause 20. There is a practice of using the contract bills to make more satisfactory provisions than the contract provides. Clause 2.2.1 reads:

> Nothing contained in the Contract Bills shall override or modify the application or interpretation of that which is contained in this Articles of Agreement, the Conditions or the Appendix.

The phrase in the pre-JCT 80 sub-clause 12(1) included the words 'or affect in any way whatsoever' after the word 'modify', and the Articles of Agreement and the Appendix have been added to the Conditions at the end of the clause.

This change amounted to an extension of the possible function of the contract bills which could include imposing on the contractor further obligations and restrictions, provided such terms do not 'override or modify' the printed wording of the contract. For example, the execution or completion of work in a specific order may be indicated in the bills, but not, for example, the requirement to complete sections of the works by dates before the completion date for the whole contract, nor to impose liquidated damages if the contractor fails to meet such dates, as this is within the area of overriding or modifying the documents mentioned (see *Gleeson* v. *Hillingdon London Borough Council* (1970)).

It is a general rule of interpretation of documents that the written or typed part overrides the printed part in the event of a clash. However, in *Gold* v. *Patman and Fotheringham Ltd* (1958) a similar clause (condition 10) in the 1939 form was considered by Hodson LJ on appeal. He said:

'The paragraph last read [in the bills], beginning with the words "Such insurances" purports to override the provisions as to insurance (save as to fire) contained in condition 15 and must therefore be disregarded, having regard to the language of condition 10. No further reference need therefore be made to this paragraph.'

There have been similar decisions concerning this clause or its equivalent which have the effect of reversing the basic rule of interpretation.

The meaning of the pre JCT 80 sub-clause 12(1) was considered in the case of *English Industrial Estates Corporation* v. *George Wimpey & Co Ltd* (1972) by the Court of Appeal. Wimpey undertook in the standard form to extend a factory owned by the plaintiff and leased to Reed Corrugated Cases Ltd. Wimpey had a blanket contractors' all risks policy which provided the cover required by sub-clause 20A(1) (now 22A.1). After a great deal of work had been carried out and Reed had installed machinery and stock in the extension, much of the new factory was destroyed by fire. At that time the contractors had not finished their work and the question was who was responsible for the fire damage (this aspect will be considered in detail in Chapter 13). However, a point which also had to be considered was whether the employers could depend upon special provisions identified as C and D inserted in the bills of quantities. These provisions showed that the parties to the contract contemplated that Reed would install machinery and equipment and would occupy and use part of the works but that, in spite of this, the contractors were still to keep the works covered by insurance. However, these provisions were at variance with clause 16 of the JCT form which provided that 'if at any time or times before practical completion of the works the employer, with the consent of the contractor, shall take possession of any part or parts of the same, then, notwithstanding anything expressed or implied elsewhere in the contract, ... such parts shall, as from the date on which the employer shall have taken possession thereof, be at the sole risk of the employer as regards any of the contingencies referred to in Clause 20A.' Thus the responsibility to insure was shifted from the contractor to the employer.

Wimpey maintained that the court could not consider provisions C and D because of sub-clause 12(1) of the agreement. Unfortunately the judges gave differing opinions in explaining the application of this sub-clause so the matter has still not been resolved.

In the court of first instance Mocatta J had said that, left to himself, he would have had regard to provisions C and D despite what sub-clause 12(1) said but he felt himself bound by observations made in previous cases. He therefore thought it best not to rely on C and D in interpreting clause 16. Lord Denning thought that the court should have regard to C and D. He said they were carefully drafted and typed in the bills of quantities. They were put in specially to enable contractors to make their calculations and tender. They were incorporated into the formal contract just as much as the JCT form conditions. In contrast, clauses 12 and 16 were printed conditions in the middle of 23 pages of small print and in general terms. On settled principles they should take second place to C and D. In his opinion the contractors remained liable to insure and could pass on the insurance premium to the employer although in fact he came to the same conclusion without placing any reliance on C and D.

Edmund Davies LJ thought provisions C and D could be used in spite of sub-clause 12(1), not to help the interpretation of the contract but to consider whether acts which might otherwise be regarded as pointing to 'taking possession' by the employer were not explicable on a different basis. When that was done, a flood of light was thrown on the course of events and underlined the judge's point that 'temporary and varying impediment to construction work differs greatly from a permanent consensual transfer of possession'.

Stephenson LJ was of the opinion that sub-clause 12(1) needed reconsideration but, as it stood, he thought that regard could not be had to paragraphs C and D, but some formality was required in applying clause 16.

All one can pick out of these opinions is that only Stephenson LJ seems freely to have adopted the literal interpretation which might not have been very practical bearing in mind the regular use of the bills by quantity surveyors to amend the contract wording despite sub-clause 12(1). Lord Denning's view was based on the principle of construction that typed and written clauses overrule the printed contract wording, but, while this rule would certainly apply to any physical alteration of the contract wording, it overlooks the positive refusal in sub-clause 12(1) to allow any modification of the contract conditions by anything contained in the contract bills.

In conclusion, there were two judges (Mocatta J in the court of first instance and Stephenson LJ in the Court of Appeal) in favour of upholding the literal meaning of sub-clause 12(1) with the other two Court of Appeal judges trying to get round the wording of this sub-clause and recognise a practice among the drawers of contract bills which was unlikely to cease. However, the three Court of Appeal judges came to the same decision, viz. that the contractors were liable for the fire loss, albeit by a different route.

Subsequent case law such as *Henry Boot* v. *Central Lancashire New Town Corporation* (1980), as well as those cases mentioned above, would seem to advise that sub-clause 2.2.1 must always be followed, but the bills could possibly be referred to, as Judge Fay did in *Henry Boot* (following Edmund Davies LJ in *Wimpey*), in order to follow what was going on. However, it must be pointed out that Judge Fay did say that the whole phrase 'quality and quantity of the work included in the Contract Sum (shall be deemed to be that which is set out in the Contract Bills)' was wide enough for him to look at the bills and find that the relevant work, although formally included, was practically excluded from the contract work. The phrase just quoted appears in the pre-JCT 80 form in sub-clause 12(1) but in JCT 80 it is the whole of sub-clause 14.1. The point is that if the query concerns the quality or quantity of the work, it is legitimate to look at the bills for guidance.

It has been suggested that, unless the clause under discussion is removed or amended, physical alteration of the conditions is the only safe way of varying the contract terms (see Stephenson LJ's judgement in *Wimpey*).

The importance of the contract wording

It may seem elementary but it is absolutely vital to ensure, when dealing with any problem concerning the meaning of the JCT contract, that not only the right edition but also the right version and amendment which apply are considered. This dire warning also applies to the wrong assumption that all types of JCT contract, e.g. Minor Works and the basic standard contract, will produce the same result for a given situation, as a difference in wording will make all the difference to the legal decision. For example, if the facts of *Ossory* exist when the JCT Minor Works contract applies a different result would appear. This is illustrated in the case of *National Trust* v. *Haden Young Ltd* (1994), which is considered when dealing with the Minor Works form in Chapter 16.

Commentators have called for clarity and simplicity of these JCT complex provisions, and recent decisions make that need more pressing. Sir Michael Latham's final report of the Government/Industry Review of Procurement and Contractual arrangements in the UK construction industry was published in the title *Constructing the Team*. The executory summary of the Report identifies recommendations, which include the following: 'Endlessly refining existing conditions of contract will not solve adversarial problems. A set of basic principles is required on which modern contracts can be based.'

Chapter 4

Insurance against injury to persons and property and excepted risks: clauses 21.1 and 21.3

JCT 80 provides for liability insurance against injury to persons and property as follows:

21.1.1.1 Without prejudice to his obligation to indemnify the Employer under clause 20 the Contractor shall take out and maintain insurance which shall comply with clause 21.1.1.2 in respect of claims arising out of his liability referred to in clauses 20.1 and 20.2.

.1.2 The insurance in respect of claims for personal injury to or the death of any person under a contract of service or apprenticeship with the Contractor, and arising out of and in the course of such person's employment, shall comply with the Employer's Liability (Compulsory Insurance) Act 1969 and any statutory orders made thereunder or any amendment or re-enactment thereof. For all other claims to which clause 21.1.1.1 applies the insurance cover:

- shall indemnify the Employer in like manner to the Contractor but only to the extent that the Contractor may be liable to indemnify the Employer under the terms of this Contract; and
- shall be not less than the sum stated in the Appendix [1.1] for any one occurrence or series of occurrences arising out of one event.

21.1.2 As and when he is reasonably required to do so by the Employer the Contractor shall send to the Architect for inspection by the Employer documentary evidence that the insurances required by clause 21.1.1.1 have been taken out and are being maintained, but at any time the Employer may (but not unreasonably or vexatiously) require to have sent to the Architect for inspection by the Employer the relevant policy or policies and the premium receipts therefor.

21.1.3 If the Contractor defaults in taking out or in maintaining insurance as provided in clause 21.1.1.1 the Employer may himself insure against any liability or expense which he may incur arising out of such default and a sum or sums equivalent to the amount paid or payable by him in respect of premiums therefor may be deducted by him from any monies due or to become due to the Contractor under this Contract or such amount may be recoverable by the Employer from the Contractor as a debt.

Clause 21 is an insurance clause as it requires insurances to be effected to support the indemnity given under clause 20 by the contractor, and employers' and public liability policies are the main insurances concerned (see Chapters 5 and 6).

4.1 Detailed conservation of the wording of clause 21.1

Introductory paragraph

> Without prejudice to his liability to indemnify the Employer under clause 20 the Contractor shall take out and maintain insurance which shall comply with clause 21.1.1.2 in respect of claims arising out of his liability referred to in clauses 20.1 and 20.2.

This introductory paragraph makes clear that the indemnity given to the employer in clauses 20.1 and 20.2 is in no way limited to the insurance which the contractor is required to take out under this clause. A major alteration not introduced by the 1986 amendments but by Amendment 4 issued in July 1987, was to remove the words 'and shall cause any sub-contractor to take out and maintain insurance'. The effect of this is considered below in the analysis of the same amendment made to clause 21.1.3.

This sub-clause 21.1.1.1 is shorter and simpler than the 1980 version.

The Employers' Liability (Compulsory Insurance) Act 1969

> The insurance in respect of claims for personal injury to or the death of any person under a contract of service or apprenticeship with the Contractor, and arising out of and in the course of such person's employment, shall comply with the Employers' Liability (Compulsory Insurance) Act 1969 and any statutory orders made thereunder or any amendment or re-enactment thereof.

This rather lengthy part of this clause makes no reference to the sub-contractor but requires the contractor to insure against claims by their employees arising out of and in the course of, their employment in accordance with the 1969 Act. This Act came into force on 1 January 1972. From that date employers were required to insure against their liability for personal injury or disease sustained by their employees in the course of their employment.

Display of certificates

Insurers are required to issue approved policies and certificates confirming the contract of insurance whenever a policy is issued or renewed. The employers were required to display the certificate (or copies of it) at each of their premises (or sites in the case of the construction industry) for the information of their employees, and produce it as required.

Section 1(3) provides that an approved policy means a policy of insurance not subject to any conditions or exceptions prohibited by regulations under the Act, and business includes a trade or profession, and includes any activity carried on by a body of persons, whether corporate or unincorporate.

Section 2(1) states that an employee means an individual who has entered into or works under a contract of service or apprenticeship with the insured whether by way of manual labour, clerical work or otherwise, whether such contract is expressed or implied, oral or in writing.

Financial limits on cover

The Employers' Liability (Compulsory Insurance) General Regulations 1971 *inter alia* impose a minimum limit of £2 million in respect of any one occurrence for which an employer is required by the Act to insure. In fact employers' liability insurers issued policies which were unlimited in the amount of indemnity.

This unlimited approach stems from the days in the early part of this century when the employers' liability policy covered the employer's liability under the Workmen's Compensation Acts. It was made possible by insurers passing on this part of the exposure to their reinsurers who in turn passed it to their retrocessionaires. The effect of the unlimited risk was therefore met by those who were rather remote from the basic exposure. This tended to mean that little attention was paid to the possible catastrophe. Also there had been cases where major workplace accidents resulted in very large employers' liability claims, e.g. the Piper Alpha tragedy. As a result when policies were renewed in 1995 a limit of indemnity was introduced. Most insurers offer a standard minimum limit of £10 million for any one occurrence. However, higher limits are available.

Prohibited conditions

The Employers' Liability (Compulsory Insurance) General Regulations also deal with the prohibition of certain conditions in policies. The conditions concerned are those that provide that no liability shall arise under the policy, or that any such liability so arising shall cease:

- In the event of some specified things being done or omitted to be done after the happening of the event giving rise to a claim under the policy, e.g. giving written notice to the insurers as soon as possible after the accident.
- Unless the policyholder takes reasonable care to protect his employees against the risk of bodily injury or disease in the course of their employment.
- Unless the policyholder complies with the requirements of any enactment for the protection of employees against the risk of bodily injury or disease in the course of their employment.
- Unless the policyholder keeps specified records or provides the insurer with or makes available to him information therefrom.

Thus insurers cannot use a breach of these conditions in order to avoid paying a claim and the above remarks will facilitate the understanding of the wording in the operative clause of the employers' liability policy under the heading 'Avoidance of certain terms and rights of recovery' (see Chapter 6).

Exemptions from the Act

Section 3 and subsequent statutory instruments exempt certain government departments, nationalised industries, public and local authorities etc. from the Act. In this connection reference should be made to the effect of the Employers' Liability (Compulsory Insurance) Exemption (Amendment) Regulations 1992 SI 1992 No 3172 in adding a fresh exemption to the list of exemptions from the employer's requirement to take out liability insurance (see Chapter 6).

Ineffective cover

However, the case of *Dunbar* v. *A & B Painters and Economic Insurance Co and Whitehouse & Co* (1985) illustrates that not all policy terms are prohibited by the Employers' Liability (Compulsory Insurance) Act 1969 from operating to prevent insurers denying liability under the employers' liability policy. The Court of Appeal held that where insurers had repudiated such a policy by reason of misrepresentation by the insurance brokers, but would in any event have been entitled under a term of the policy to avoid liability to the insured employer, the court, on the insured's claim against the brokers for damages for loss of indemnity under the policy, should assess the likelihood that the insurers would have avoided liability and reduce the insured's damages accordingly.

This was an appeal by the second third party, the insurance brokers, from an order that they indemnify the defendant insured, A & B Painters, in respect of its liability to pay the plaintiff damages for injuries sustained when he fell over 40 feet in the course of his employment with the defendant. The term of the policy entitling the insurers to refuse to pay the plaintiff was that it did not cover work in excess of 40 feet in height. In the course of his judgment May LJ stated that the judge at first instance should have assessed the chance of Economic succeeding in taking the height point and concluded it was nil. Whether the judge applied this test or the balance of probabilities approach is not clear but the conclusion was the same. Had he applied the percentage test indicated by the Court of Appeal, he might have decided that there must have been some remote chance that Economic might have taken the point successfully. However, the appeal failed.

Balcombe LJ 'shared the deep concern' which the trial judge expressed at this repudiation of an employers' liability insurance which appeared to drive a coach and horses through the provisions for compulsory insurance in the 1969 Act and Regulations which required the certificate of insurance to be displayed by the insured so that employees could inspect it. The certificate was said to be 'a snare and a delusion' as it misled employees into believing adequate cover existed, when the fact was that no effective insurance was in force in this case, either because of misrepresentation or because of the height warranty.

The truth of the matter is that the 1969 Act does not provide employees with the same amount of protection that victims of motor accidents enjoy through the Road Traffic Act restrictions on the avoidance of liability and the support of the Motor Insurers' Bureau. While it may be avoiding the issue to say that, if an employers' liability policy is ineffective the employer may have sufficient money to satisfy the judgment debt, the likelihood of the Government establishing an Employers' Liability Insurance Bureau is remote, unless sufficient cases of this nature arise, which is also unlikely.

This matter was re-aired, however, in *Richardson* v. *Pitt-Stanley and Others* (1994). Here, when it transpired that the company employing the injured plaintiff was uninsured, he sought to impose personal liability on the directors of the company, it being alleged that the company's failure to insure was a breach of statutory duty and that the directors were responsible in tort for the company's omission. The authority for this allegation was *Monk* v. *Warbey* (1935) where it was held that the owner of a vehicle allowing a person to use that vehicle without insurance is liable for the tort of breach of statutory duty for causing or permitting the uninsured use of the vehicle. It was argued that the 1969 Act should be interpreted in the same way. The Court of Appeal construed the legislation as intended only to impose criminal and not civil liability, and the directors could only be fined, not sued, for damages. There was, however, a dissenting judgment.

Unlimited liability under the contract and in common law

For all other claims to which clause 21.1.1.1 applies the insurance cover:

- shall indemnify the Employer in like manner to the Contractor but only to the extent that the Contractor may be liable to indemnify the Employer under the terms of this Contract; and
- shall be not less than the sum stated in the Appendix [1.1] for any one occurrence or series of occurrences arising out of one event.

This last sentence of sub-clause 21.1.1.2 refers back to the public liability policy required by sub-clause 21.1.1.1. The minimum limit of indemnity required for any one occurrence or series of occurrences arising out of one event is stated in the Appendix, but the policy issued will be unlimited in respect of any one period of insurance. It is important to appreciate that the contractor's liability under the contract is, however, unlimited in respect of any one occurrence or series of occurrences arising out of one event as this is the indemnity he gives to the employer under clause 20, which clause 21 is stated not to prejudice (in the first line of clause 21.1.1.1).

Furthermore, common law liabilities for personal injuries to and damage to the property of third parties are unlimited. Thus it is possible that if too low a limit of indemnity is chosen in the Appendix, the contractor could find himself with a greater liability under the contract than the limit of indemnity under his public liability policy, even though he has complied with the contract conditions. This in turn could rebound on the employer if the contractor is unable to meet his financial responsibilities.

Incidentally the insurances referred to are any insurances which provide cover for personal injury or death or for injury or damage to property so they are not limited to employer's liability and public liability policies. Thus it could include motor insurances. Furthermore, as the minimum cover for any one occurrence or series of occurrences set out in the Appendix applies to all such policies, it is important to see that the commercial motor insurance policy of the contractor, which has a limit of indemnity for damage to property, complies with the contract requirements. The contractor may insure for a sum (as a limit of indemnity) greater than that stated in the Appendix. This is stated in the contract by a footnote in square brackets.

Professional responsibility for checking the insurance

Who does the checking?

> As and when he is reasonably required to do so by the Employer the Contractor shall send to the Architect for inspection by the Employer documentary evidence that the insurances required by clause 21.1.1.1 have been taken out and are being maintained, but at any time the Employer may (but not unreasonably or vexatiously) require to have sent to the Architect for inspection by the Employer the relevant policy or policies and the premium receipts therefor.

It is interesting to note how the responsibility for checking insurance cover has varied under the JCT contract over the last fifteen years or so, bearing in mind the tendency of architects and other construction professionals to use the edition with which they are familiar whether it is the current one or not. Prior to the 1980 edition clause 19(1)(b) (dealing with the insurance against injury to persons and property) required the contractor, when required to do so by the architect, to produce *for inspection by the employer* documentary evidence that the insurances required were properly maintained and when required produce for inspection the policies and receipts in question. Clause 20A concerning insurance of the works uses the same wording. The 1980 Edition in clause 21, which took the place of clause 19 of the previous edition, requires the contractor, when required by the architect, to produce *for inspection by the architect* the insurance evidence etc. The insurance of the works, clause 22A, uses the same wording. In the 1980 Edition following the 1986 amendment, the equivalent clauses make the employer responsible for inspection of these insurances. Consequently it depends on the edition and revision or amendment used as to whether the architect or employer is expressly responsible for inspection of the insurance required.

In *Samuel Smith Old Brewery (Tadcaster) and Others* v. *G S Cronk Builders (Tinten) Limited and Alan Miles* (1982) the architect was sued by the employer under the pre-1980 Edition of the JCT contract for inadequate design and supervision, and also for negligently approving the insurance documents purported to be produced by the contractor under clause 19(1)(b). The contractor, it was alleged, never obtained the required insurances set out in clause 19(1)(a). Consequently, it was alleged, the losses incurred by the plaintiffs were not recoverable under any insurance policy. Without going into further detail (other than to say that the plaintiffs only succeeded against the builder) it is clear that whether the construction contract requires the professional, instructed by the employer, to inspect the insurance documents, or merely to be the intermediary in obtaining that evidence (but the latter assumes that inspection responsibility as a volunteer) there is always a danger that the employer will blame the professional for any insurance discrepancy. So it does not matter that the construction professional is not a party to the construction contract (although he is a party to his Conditions of Engagement), nor does it matter that he is not expressly responsible for inspection of the contractor's insurances. The point is whether he voluntarily accepts that responsibility; it appears that he often does so as alleged in *Samuel Smith.*

In the Annexes to the Likierman Report on Professional Negligence the Construction Study Team in Annex B give a Summary of the Capper/Uff Report and on page 42 the following comment concerning advisory work is made:

> 'Initial advisory and planning services frequently extend to areas outside the traditional role of construction industry professionals. For example, they may involve consideration of financial or legal matters. In regard to such work, the law will not excuse the professional on the ground that the area of expertise was outside his or her professional competence (*Morest Cleaners* v. *Hicks* (1966) 2 Lloyd's Rep 338; see also *BL Holdings* v. *Wood* 12 BLR 1). Where expertise in other fields is required, the professional should either employ a specialist directly or advise clients to employ specialists themselves.'

This comment would clearly apply to the inspection of insurance policies but it is doubtful whether the advice in the last sentence is taken. In any event unless the contractor's insurance broker is used (and even he may require a fee) the question arises as to who is going to pay the insurance specialist's fee?

The evidence to be produced

The first part of this sub-clause 21.1.2 poses the question as to what is meant by *documentary evidence* because it seems from the second part that it does not include the policy or policies and premium receipts. The earlier form of the 1963 JCT contract accepted a current certificate of insurance in place of the policy or policies and premium receipts. Now whether this was abandoned in the later 1963 edition to avoid confusion with the employers' liability certificate of insurance required by the 1969 Act mentioned earlier, or whether it was due to the criticism with which the certificate idea met, is not known. It was criticised because a certificate would have to indicate policy exclusions, otherwise it is of little use, and thus could be a lengthy document and troublesome to prepare, and even then not be entirely accurate.

Consequently a certificate may not achieve its object as it is easier, more accurate and quicker for all parties concerned to produce the policies and premium receipts.

On occasions a contractor may find that his annual liability policies have to be produced at the same time to different employers. In such cases duplicate policies and premium receipts would be more satisfactory than a detailed current certificate of insurance. It is of little help to use the words 'subject to the usual terms and conditions of the policy' in a certificate as there is no standard public liability policy in the insurance market and such a reference means nothing to the building owner and those who act for him without reading the policy concerned.

Since *Hedley Byrne & Co Ltd* v. *Heller and Partners* (1964) it is possible that the issue of a certificate of insurance may carry with it a liability to the employer for errors or omissions concerning the details of the insurance stated in the certificate. However, it is presumed that the insurer could contract out of any such liability. Nevertheless this disclaimer of liability would detract from the value of any certificate. In these days when all insurance offices have a copying machine it does not entail much extra work to produce a duplicate policy and current premium receipt and this is another reason in favour of the latter action.

It is difficult to suggest any alternative documentary evidence (to the policies and premium receipts) that the insurances are properly maintained, to which the first part of this sub-clause could refer.

The right of the architect to call for the actual policy or policies and receipts must not be exercised *unreasonably or vexatiously*. Presumably it would not be unreasonable to call for a sight of these documents when a contract has been awarded to the contractor and they have not been seen by the employer since the last annual renewal date. Also it would be reasonable for the employer to make a further request for the documents after renewal when renewal falls due during the period of a contract.

Responsibility for insurance

> 21.1.3 If the Contractor defaults in taking out or in maintaining insurance as provided in clause 21.1.1.1 the Employer may himself insure against any liability or expense which he may incur arising out of such default and a sum or sums equivalent to the amount paid or payable by him in respect of the premiums therefor may be deducted by him from any monies due or to become due to the Contractor under this Contract or such amount may be recoverable by the Employer from the Contractor as a debt.

By Amendment 4 referred to earlier the words 'and shall cause any sub-contractor to take out and maintain insurance' were removed from the introductory clause 21.1.1.1 in 1987 and in this sub-clause also the contractor's responsibility in the previous version to see that the sub-contractor insures is omitted. It had always been doubted whether the employer had an insurable interest (see the second principle of insurance under the heading 'Insurance law' in Chapter 2) which would entitle him to effect an insurance which a *sub-contractor* has failed to effect. For example, if a sub-contractor, to whom the contractor has sub-let part of the work, fails to insure his liability to third parties, this might lead to the sub-contractor's bankruptcy, but the consequent delay would be the responsibility of the main contractor and it is argued that the interest of the employer is remote. However, on the other hand the bankruptcy of the contractor, or even the sub-contractor, can eventually cause a loss to the employer, and so he has a right to prevent that loss. In any event until now contractual liability gave an insurable interest. The meaning of 'defaults in taking out or in maintaining insurance' is considered in more detail at the end of this chapter under the heading 'General matters concerning clause 21', but failure to take out insurance involves two interpretations. It means either that the employer is given authority as agent to insure for the contractor, or the employer is to effect the necessary insurance in his own name. In the former event the employer may be in some difficulty in taking advantage of his apparent authority under the sub-clause as the insurers would want some information about the contractor (particularly concerning the claims experience) before issuing cover which the employer probably could not provide. In the latter case (the employer insuring in his own name), in the event of a loss occurring for which the contractor is liable under the contract, the insurer having paid the claim would be entitled to sue the contractor by virtue of subrogation.

Regarding the recovery from the contractor as a debt presumably covers the rare situation where the employer does not owe sufficient money to the contractor to deduct the required premium.

4.2 *Excepted risks – clause 21.3*

21.3 Notwithstanding the provisions of clauses 20.1, 20.2 and 21.1.1, the Contractor shall not be liable either to indemnify the Employer or to insure against any personal injury to or the death of any person or any damage, loss or injury caused to the Works or Site Materials, work executed, the site, or any property, by the effect of an Excepted Risk.

In sub-clause 21.3 it is satisfying to note that attention has been paid to criticism in the first edition of this book of the previous wording, which did not include personal or bodily injury when excluding nuclear risks and sonic waves from the contractor's liability and his insurance requirements. This has now been done. Also, quite correctly, the old sub-clause 21.3 has been amended as it now has its own exclusion of nuclear risks and sonic waves defined in the contract in clause 1.3 as 'Excepted Risks'.

Clause 1.3 is an insurance market clause, i.e. its wording is used throughout the insurance world in the UK. It reads as follows:

> ... ionising radiations or contamination by radioactivity from any nuclear fuel or from any nuclear waste from the combustion of nuclear fuel, radioactive toxic explosive or other hazardous properties of any explosive nuclear assembly or nuclear component thereof, pressure waves caused by aircraft or other aerial devices travelling at sonic or supersonic speeds.

This sub-clause 21.3 is taken out of turn as it applies to sub-clauses 20.2 and 21.1 as well as to sub-clause 21.2, which has not yet been considered. The effect is that certain risks are specifically excluded from the indemnity given to the employer and the insurance requirements, namely:

- the effect of ionising radiations;
- the contamination by radioactivity from any nuclear fuel, waste or explosion;
- pressure waves caused by aerial devices travelling at sonic or supersonic speeds.

Thus the risk of these occurrences rests with the employer.

The tendency is for standard form contracts to ask for exclusionless policies and this cannot be obtained, although the employers' liability policy nearly reaches this ideal. The position has been recognised to some extent by this sub-clause and the definition of all risks insurance in clause 22.2, plus the exceptions to sub-clause 21.2.1 which are dealt with later. However, there are still some areas where the insurance policies fall short of the contract and insurance requirements. These areas are set out at the end of Chapter 6, not that a contract can cater for all the exclusions which may appeal to the underwriter of the contractor's policies.

However, the common exclusions should always appear in the contract, and as mentioned some headway is being made in this direction. The object should be to avoid the employer contracting for and the contractor agreeing to provide the unobtainable. The summary at the end of Chapter 6 also sets out the risks which the employer faces which are probably uninsurable.

The Nuclear Installations Acts 1965 and 1969 together with the Energy Act 1983 makes the operator of a nuclear installation solely liable for injury to any person or damage to any third party property. Only the operator is liable, notwithstanding that other parties such as contractors may be liable in tort for the consequences of a nuclear accident on the licensed site. Because the operator alone is liable, he only is required to provide insurance, and the United Kingdom Pool policy meets the operator's nuclear legal requirements. The policy is a type of liability insurance written through the British Insurance (Atomic Energy) Committee and, *inter alia*, covers the operator's nuclear liabilities under the statutes mentioned for an overall indemnity figure of £140 million (in relevant cases £20 million) for the current 'cover period'. Above £140 million the government is responsible. These figures apply from April 1994. The cover period in practice means the lifetime of a licensed installation subject to certain qualifications.

Both the public liability policy and the employers' liability policy (the latter only in respect of the indemnities given by the insured to any principal) are subject to the exclusion of nuclear risks. However, radioisotopes, which are radioactive forms of well-known substances whose properties are on occasions used by the construction industry, e.g. to test welding, are not covered by this exclusion. Thus they are covered by the insurance market.

A sonic waves exclusion has never operated in the employers' or public liability policies. It was understood that if Concorde's test flights showed that there was a real risk of claims arising under certain liability policies, there could be a reassessment of the position. However, this has not become necessary. In fact, although it is the practice to exclude this risk from material damage policies, it is doubtful whether this is justified as it has not proved to be a hazardous risk.

4.3 General matters concerning clause 21

The meaning of 'defaults in taking out and maintaining insurance'

It is assumed that this phrase means failure to insure in accordance with the contract *as far as is reasonably possible*, and not only failure to insure at all, as in practice this is how it operates. See at the end of Chapter 6 a summary of risks usually remaining uninsured in spite of the contract requirements. Nevertheless, as the employer and the architect are unlikely to be professionally competent from an insurance viewpoint, it would be advisable for them to obtain professional insurance advice as to the adequacy of the cover provided by the policies submitted to them.

A court might well decide that this would be a reasonable way for the employer and the architect to discharge their rights or responsibilities under the contract, namely for the employer to insure if this is not done and for the architect to examine the contractor's policies. In practice, if the cover is considered to be inadequate, bearing in mind the cover normally available, the policies are usually returned to the contractor with the request to rectify the cover as required. In this connection, risks can arise out of the use of motor vehicles and mechanically propelled mechanical plant or vessels or boats which are excluded from the public liability policy. In these cases enquiries should be made of the contractor as to the liability cover provided by his motor and marine policies, assuming these apply. On rare occasions aircraft may be used and then an aviation policy would have to be produced. The point is that sub-clause 21.1 does not specify policies by name but the insurances are as set out in sub-clauses 21.1.1.1 and 21.1.1.2.

It is a pity that, having set out an operative clause and exceptions for a contractors' all risks policy in clause 22.2 in the 1986 amendments, the JCT never extended this approach to the public liability policy, or for that matter to the employers' liability policy, although in the latter the exceptions are really non-existent. Those responsible for inspecting the contractor's insurances would appreciate such an approach.

The period of insurance

Neither clause 20 nor clause 21 state the period for which the insurances should be maintained. As the liability of the contractor to indemnify the employer could apply well after the completion of the contract (because under the Limitation Act 1980 the cause of action accrues when the employer has suffered loss and this occurs when judgment is delivered against him), it behoves the contractor to keep his liability policies in force. In fact, as the employers' and public liability policies are renewed annually and not contract by contract, there is little danger of a contractor allowing these policies to lapse.

However, a contractor might change his insurer, and then the insurer handling the claim will be the one covering the policy period during which the loss occurred. Occasionally, this can result in the difficulty of discovering when the loss occurred, and thus deciding which insurer should handle the claim. Unfortunately, experts in the shape of engineers or architects do not always agree on when property damage occurs. That is when insurers begin to wonder whether their policies should have been written on a claims-made basis, i.e. the policy operating if the *claim is made* during the period of the policy, which is a much easier date to decide, thus avoiding this dispute between insurers. This may appear to be another story, but it should be appreciated that in any dispute between insurers it is the insured who suffers because of the delay that occurs while the insurers decide who is to handle the claim. It is worth noting that, as a result of *Pirelli General Cable Works Ltd* v. *Oscar Faber & Partners* (1982) which decided that in the case of damage to property in a tort action the cause of action accrues when the damage occurs, lawyers are in the same position in deciding, in a tort case, when the damage occurred as insurers can be under their policies in deciding which insurer is to handle a claim.

Sometimes the public liability cover is given as part of the contractor's all risks policy covering the works. There is then a danger of the public liability cover lapsing with the contractor's all risks policy. In such cases it would be advisable for the contractor to have a contingency cover or policy to take care of the continuing liability. Contractors' all risks policies are not always taken out contract by contract, but mainly the standard policy covers the period of a particular contract.

The indemnity to the employer

An important aspect of clause 21.1 is that it does not require the insurances mentioned to be in the joint names of the contractor and the employer, nor does it require a direct indemnity to the employer as principal. Thus the employer cannot call directly on the contractor's insurers to indemnify him in respect of any third party claim. The employer must claim under the JCT contract against the contractor for an indemnity and the contractor will then request his insurers to handle the claim in accordance with the liability the contractor has assumed under the JCT form.

Chapter 5

Classes of policy providing the cover required

After each insurance clause of the contract has been discussed, it is logical to explain the policies that provide the cover required by that clause. It is also logical, having considered the insurance policies in classes in this chapter, to follow with an explanation of the individual policies, but it is not possible to achieve both these ideals.

Consequently a compromise has been decided upon which to some extent achieves the best of both alternatives. Thus where a contract insurance clause has been considered in detail, the next chapter as far as possible will discuss the policy that attempts to provide the cover required. This leaves out the contractors' all risks policy, which, although called for by clause 22, will be considered immediately after the chapter on the liability policies as an illustration of a material damage policy in contrast to a liability policy, these being the main classes of policies to be considered by this book.

Clause 22 has three optional clauses and optional clause 22C has two sub-clauses 22C.1 and 22C.2 and the former, concerning existing structures and their contents, only requires cover for the specified perils (not a contractors' all risks policy). This policy in the insurance world is known as a fire and special perils policy. The majority of clause 22 concerns responsibility for the works etc., so here a contractors' all risks policy is required as clause 2.1 states that the contractor shall carry out and complete the works. 'All Risks Insurance' is defined in clause 22.2 and 'Specified Perils' in clause 1.3.

The case of *Charon (Finchley) Ltd* v. *Singer Sewing Machine Co Ltd* (1968) illustrates the position, although not a case under the JCT contract. Here the defendants employed the plaintiffs to convert premises into a shop. Hooligans broke into the premises and caused damage. At that time work on the shop was complete except for fixing mosaic in the display windows. Singer's surveyors instructed Charon to make good the damage so that the shop could open at the earliest opportunity. The making good involved doing again work already done and paid for under interim payments made on the surveyor's certificate. No arrangement was made as to payment for this extra work and the plaintiffs sued for this sum.

The court held that the original contract was an entire contract. Consequently, before full payment had to be made by Singer, they were entitled to have the work completed. Since the contract was not completed, the obligation remained on Charon to complete it, and if that involved doing again work which had been damaged, then that was a liability the builder had to shoulder, even though it was wilful. The court quoted *Hudson's Building and Engineering Contracts*, 9th edition, page 223:

> 'Indeed, by virtue of the express undertaking to complete (and in many contracts to maintain for a fixed period after completion) the contractor would be liable to carry out his work again free of charge in the event of some accidental damage occurring before completion even in the absence of any express provision for protection of the risk.'

5.1 The liability policies

The two liability policies concerned are those that insure the liability of the employer (employer here is used in the sense of the master in the master-and-servant relationship) (a) to his own employees and (b) to the public. The third party section of a third policy (namely the motor policy) is involved under (b), but for convenience will be considered below when discussing the exclusion of mechanically propelled vehicles under the public liability policy.

The cover provided by these policies will be considered in the next chapter, but there is another liability policy which results directly from the JCT form, namely that required by sub-clause 21.2.1. This is considered in Chapter 10.

An insurance policy does not cover every risk and the various reasons for this are given when the exclusions and limitations of the policies are considered later in this book.

Policies usually contain a note in a prominent position that they should be read and returned for correction if found to be incorrect. However, policies are not often examined by a policyholder until a loss occurs or a contract such as the JCT form requires such an examination. Insurers are naturally anxious to help the policyholder understand the policy cover, but it is for him to approach the insurer or his broker on any aspect that is not clear to him. Therefore an insured should read the whole policy, as the cover can sometimes be widened by deleting certain exclusions or increasing the limit of indemnity (or sum insured in the case of a material damage policy) on payment of an additional premium. The exclusions should therefore be read particularly carefully. Incidentally, the remarks on this and the previous paragraph apply equally to a material damage policy.

It is inevitable when considering the extent to which underwriters are prepared to delete standard exclusions or exceptions that comments can only be made in a general way. Underwriters vary in their opinions of the seriousness of certain risks, and it is a good thing that this is so otherwise progress in providing the required cover would not be forthcoming. For example, at one time the subsidence risk under the public liability policy was almost universally considered uninsurable on an open basis, i.e. cover was only undertaken contract by contract, and even a decade ago a considerable number of policies contained a lower limit of indemnity for the subsidence risk than was available for the other risks covered by the policy. Nowadays it is unusual to find subsidence not covered up to the same limit of indemnity as the remainder of the policy cover.

The position can be summarised as follows.

Employers' liability

This policy basically covers the liability of an employer (master) to his employees (those persons under a contract of service or apprenticeship with the employer) for bodily injury or disease arising out of and in the course of their employment.

Public liability

This policy basically provides an indemnity against personal injury claims by the public (other than employees), and property damage claims.

Third party section of the motor policy

Dealing with car insurance first, the cover is not confined to accidents on roads as defined in the Road Traffic Acts, but applies equally to accidents on private property. As well as covering the policyholder and permitted drivers against their liability to third parties, including passengers, for death and bodily injury, the third party cover provides protection against the following risks:

- property damage;
- indemnity to passengers while using the car;
- indemnity to employer or partner;
- legal costs.

The main points of difference between the normal third party goods-carrying vehicle and the third party car insurance are that the indemnity under the latter (but not under the automatic continental use extension) is unlimited in amount both for personal injury, including passengers, and for damage to property. In the case of commercial vehicles, while the indemnity for personal injury is unlimited, that for property damage usually carries a maximum indemnity limit in the region of £2 million, but higher limits may be available for an additional premium.

Clause 21.2.1 policy

This policy is named after a clause in the JCT 80 form which originated in the 1963 Edition and was then numbered 19(2)(a). This requires the contractor to protect the employer (the building owner) only, and it is not an insurance to protect the contractor. It is a hybrid policy in that it mainly gives liability cover but it also contains an element of material damage cover.

This insurance protects the employer against damage to any property owned by him (which is the object of the material damage policy) as well as damage to third party property (which is the area governed by the liability policy) arising out of the carrying out of the works (but excluding damage to the works) as a result of collapse and subsidence perils. The policy operative clause wording follows the contract clause 21.2.1 and is very wide in that it refers to 'any expense, liability, loss, claim or proceedings', but the exceptions are very restrictive. Full details are given in Chapter 10.

Professional indemnity

This policy protects the insured against his legal liability to pay damages to third parties who have sustained injury, loss or damage due to the negligence of the insured or his staff in the professional conduct of the business, i.e. liability is based on breach of professional duty which is the failure to exercise a degree of care and skill ordinarily exercised by members in the profession concerned. More detail will be given in Chapter 17.

5.2 *Material damage policies*

The type of insurance that comes under this heading covers loss or damage to property in which the insured has an insurable interest which may arise through ownership, possession or contact. A material damage policy does not cover the insured's legal liability for injury to persons or damage to third party property in which he has no insurable interest other than the legal liability which is potential and exists in respect of all persons and property the insured, as a wrongdoer, may injure or damage. In view of the title of this book, the material damage policies discussed are the contractors' all risks policy, the fire and special perils policy, and to some extent the insurance of the contractor's plant and equipment.

The two main policies that insure the works are the contractors' all risks policy (usually referred to as the CAR policy), and the fire and special perils.

The CAR policy

This policy indemnifies the insured for loss, damage or destruction of any of the property specified in the schedule while on the contract site for which the insured is responsible in accordance with the terms of the contract arising from any cause whatsoever, subject to the terms and conditions of the policy (see clause 22.2). Apart from clause 22.2 most policies will cover materials in transit and off-site goods certified and paid for (see Chapter 7 and clauses 16 and 30.3).

The fire and special perils policy

This policy gives more limited protection than the CAR policy as the intention is to cover only those perils listed as 'specified perils' in the JCT contract, subject to the policy terms, exceptions and conditions (see clause 1.3).

Insurance of contractor's plant and equipment

This insurance of loss or damage by the contractor can be arranged under a separate contractors' plant policy, but cover can also be given as an extension of the following policies:

- a CAR policy;
- a fire and special perils policy;
- an engineering policy; and
- a comprehensive commercial vehicle policy.

5.3 *Composite and combined policies*

These policies cover in one document risks which are normally covered by separate policies. In the case of the contractors' combined policy, the risks covered are legal liability and material damage. A specimen policy will be found in Appendix 2 where it will be seen that section 1 of that policy covers the employers' liability risk, Section 2 the public liability risk, Section 3 covers loss or damage to the contract works (which is the cover provided by the CAR policy mentioned earlier) and Section 4 the 21.2.1 risks.

These composite or combined policies have an advantage over separate policies because only one proposal form is required, one policy is issued, and (assuming an annual basis) one renewal notice and one premium suffice.

Another type of combined policy concerns liability risks only, but it goes a stage further than the composite policy just described as, although the cover is confined to the employers' and public liability risks and does not include loss or damage to the contract works, it does not separate the liability risks into sections with their own exclusions but contains a combined operative clause covering bodily injury to and damage to the property of any third party (employee or member of the public). However, in this type of policy it is necessary to indicate where an exception applies only to personal injury or only to damage to property, or does not apply to an employee of the person indemnified.

5.4 *Project insurance*

In the case of large construction projects the owner who commissions the project often wastes money as a result of the system of making every participant arrange separate insurance policies for his part of the work, his own plant and equipment, and for the liabilities to workmen and other third parties which arise from performance of his part of the work.

Comprehensive project (sometimes referred to as 'wrap-up' or 'omnibus') insurance, preferably arranged by the owner, has frequently been advocated and the arguments in its favour in theory seem overwhelming. Yet it is not always done, and even when arranged it is often confined to the conventional risks only, i.e. loss or damage to the works and to other property on the site and liability to third parties, including cross liabilities between the parties to the joint names policy.

As this is an important subject to the construction industry, a separate chapter is devoted to it and the pros and cons are considered in Chapter 8. It is sufficient to say here that one of the main obstacles to 'total project' insurance is the difficulty of including in the cover the professional negligence risks of the architect, engineer and quantity surveyor. The design and supervising risks of the architect and engineer as well as the design risks of the main contractor and specialist sub-contractors have proved to be very hazardous. More will be said on this subject in Chapter 17 when the 'design and build' contract forms are considered.

Chapter 6

Employers' and public liability policies

6.1 The employers' liability policy

As explained in Chapter 5, this policy covers the insured's legal liability as employer to any person who is under a contract of service or apprenticeship with the insured for bodily injury or disease arising out of and in the course of employment by the insured in connection with the business described in the policy schedule. In Appendix 2 of this book an example of this policy is given in Section 1.

Because it is compulsory for employers to insure their liability to their employees by statute, as explained in Chapter 4, and because of the restrictions placed on insurers by the regulations issued under the 1969 Act also mentioned in Chapter 4, insurers have found it necessary to make it clear to their insureds that, having paid damages to an employee under the policy in circumstances where the insurer would not have been liable to pay but for the provisions of the compulsory insurance law, they have a right of recovery against the insured. The clause of the policy setting out this position will be found in Appendix 2 under the heading in Section 1 'Avoidance of Certain Terms and Right of Recovery'.

Farrell v. *Federated Employers' Insurance Association Ltd* (1970), which was heard before the 1969 Act came into operation, is an illustration of a repudiation by the defendants that was justified but could not now result in the employee failing to obtain his damages. Farrell obtained judgment and costs against his employers, but the judgment remained unpaid as the employers were in liquidation. He claimed directly against the insurers under the Third Parties (Rights Against Insurers) Act 1930. The defendants disputed the claim on the ground, *inter alia*, that they had no knowledge that the proceedings against the employer had begun.

It was held that an omission to notify or forward to an insurer a writ served on an employer in a case of industrial injury resulted in prejudice against the insurer in this case. They had lost the opportunity of causing an appearance to be entered. It was no answer to say that the insurers might have had the judgment set aside if they had applied to the court in time. Being put to the necessity of applying to have a judgment set aside was prejudice enough if prejudice must be shown.

It should be appreciated that a policy condition required every writ served on the employers to be notified or forwarded to the insurers, and the 1930 Act did not provide for any variation in the conditions of the policy in favour of the third party (the employee in this case). As a result of the 1969 Act Regulations, an insurer cannot repudiate policy liability 'in the event of some specified things being done or omitted to be done after the happening of the event giving rise to a claim under the policy'. There are other similar prohibitions on repudiations of policy liability by insurers. However, the object is to protect the innocent claimant, thus the insurer is quite justified in these circumstances to attempt a recovery from the insured who is in breach of the policy conditions. In fact the 1969 Act Regulations say so. Further, note the limitations of the 1969 Act illustrated in the cases of *Dunbar* and *Richardson* in Chapter 4.

The recital and operative clause

The recital clause in Appendix 2 reads:

> In consideration of the payment of the premium the Independent Insurance Company Limited (the Company) will indemnify the Insured in the terms of this Policy against the events set out in the Sections operative (specified in the Schedule) and occurring in connection with the Business during the Period of Insurance or any subsequent period for which the Company agrees to accept payment of premium.

> The Proposal made by the Insured is the basis of and forms part of this Policy.

The premium is the consideration for which the insurers issue the policy. The 'Insured', 'Business', 'premium' and 'Period of Insurance' are all set out in the policy schedule (see Appendix 2). The insured events must arise from the business and occur in the insurance period.

'Proposal' appears in the 'Definitions' Section as does the extensive definition of 'Business' as follows:

> **Proposal** shall mean any information provided by the Insured in connection with this insurance and any declaration made in connection therewith.

> **Business** shall include
> (a) the provision and management of canteens clubs sports athletics social and welfare organisations for the benefit of the Insured's Employees
> (b) the ownership repair maintenance and decoration of the Insured's premises and the provision and management of first aid fire and ambulance services
> (c) private work carried out by an Employee of the Insured (with the consent of the Insured) for any director partner or senior official of the insured.

The operative clause

Omitting the paragraph headed 'Avoidance of Certain Terms and Right of

Recovery', which has already been explained earlier, the operative clause or insuring clause reads:

> In the event of Bodily Injury caused to an Employee within the Territorial Limits the Company will indemnify the Insured in respect of all sums which the Insured shall be legally liable to pay as compensation for such Bodily Injury arising out of such event.

World-wide

The indemnity granted by this Section extends to include liability for Bodily Injury caused to an Employee whilst temporarily engaged in manual work outside the Territorial Limits

Provided that

(a) such Employee is ordinarily resident within Great Britain Northern Ireland the Isle of Man or the Channel Islands
(b) the Company shall not be liable to indemnify the Insured in respect of any amount payable under Workmen's Compensation Social Security or Health Insurance legislation.

'Bodily injury' and 'Employee' are defined as follows:

Bodily Injury shall include

(a) death illness or disease
(b) wrongful arrest wrongful detention false imprisonment or malicious prosecution
(c) mental injury mental anguish or shock but not defamation.

Employees shall mean

(a) any person under a contract of service or apprenticeship with the Insured
(b) (i) any labour master or labour only sub-contractor or person supplied or employed by them
   (ii) any self-employed person
   (iii) any person hired or borrowed by the Insured from another employer under an agreement by which the person is deemed to be employed by the Insured
   (iv) any student or person undertaking work for the Insured under a work experience or similar scheme

while engaged in the course of the Business.

Employee or sub-contractor

The basic cover is for the insured's liability to employees under a contract of service or apprenticeship with the insured. However, it is the right of control which is the main aspect which decides whether the relationship is that of employer/employee or employer/independent contractor. The latter relationship is not covered by the employers' liability policy, except as mentioned below.

Self-employed contractors and labour gangs can give rise to doubts as to their legal relationship with the employer (the insured), i.e. whether they are employees or independent contractors. Therefore it is the general practice of insurers to define them as employees in order to make it clear that the employers' liability policy will handle claims made against the insured by such persons, provided that payments made by the insured to them as wages or fees are declared for the purposes of calculating the premium. Nevertheless, this still allows the insurer to defend any claim against the insured if it can be argued legally that such persons are not employees and by doing so show that, for example, no duty of care is owed to that person by his employer in the particular circumstances.

The same concession of cover applies to employees of another person who are lent to the insured on the basis that they are employees of the insured. In fact an employee's contract of service is personal to him and only he can agree to change it so that the position of such people is similar to that of the self-employed persons under the employers' liability policy. Their claims are handled under that policy but their legal status remains as it was.

Students and persons undertaking work for the insured under a work experience or similar scheme are usually employees in law and this paragraph merely makes the position clear.

Lane v. *Shire Roofing Co (Oxford)* (1995) is a typical example of the position. The plaintiff was a builder and roofer who had worked as a one-man firm for several years and was self-employed for tax purposes, but started working for others when work became scarce. The defendant, who did not take on employees but instead hired workmen for individual jobs, asked the plaintiff to undertake a re-roofing job for an all-in fee. While working on this job, for which he provided no materials but used his own ladder and tools, he fell from the ladder and sustained injuries.

A High Court judge decided that the defendant was not liable for those injuries because the plaintiff was working as an independent contractor at the time. The Court of Appeal allowing the appeal, stated that while there were distinct benefits for employers and workers in avoiding the label of employer or employee, it was important to make sure that legal responsibility for safety at work was correctly apportioned. Various factors could be taken into account in determining whether an employer/employee relationship existed, including the test of who had control and whose business it was. In the present case, it was held that the plaintiff was neither an employee nor a specialist sub-contractor, but was nearer the former because the business was clearly the defendant's business. The defendant was in breach of statutory duty as employer but there was contributory negligence by the plaintiff.

See Appendix 11 for the common law position regarding liability for sub-contractors.

Breadth of cover

The word 'Business' as mentioned in the recital clause and in the definition of 'Territorial limits' below is wide. Canteen and club administration, and management of first aid, fire and ambulance services carries a liability to the employees involved which is covered by the policy. Decoration of the insured's property is clearly within the insured's business as a construction contractor but private work carried out for a senior official of the insured would be considered a different activity, nevertheless it comes within the cover provided by the employers' liability policy.

Territorial Limits shall mean
(a) Great Britain Northern Ireland the Isle of Man the Channel Islands or off shore installations within the continental shelf around those countries
(b) member countries of the European Economic Community where the Insured or directors partners or Employees of the Insured who are ordinarily resident in (a) above are temporarily engaged on the Business of the Insured
(c) elsewhere in the world where the Insured or directors partners or Employees of the Insured who are ordinarily resident in (a) above are on a temporary visit for the purpose of non-manual work on the Business of the Insured.

Some insurers exclude temporary visits to the United States and Canada.

Time limits

Legal actions concerning personal injury must usually be commenced within three years of the date of accident. However, the law allows exceptions because some diseases take a time to develop and may not become obvious for several years, possibly even after the policy has lapsed, e.g. the insured changes his insurer. The Limitation Act 1980 caters for this situation to a considerable extent by allowing (in certain circumstances) an employee to bring an action although more than three years have passed since the accident date. Examples in this situation are deafness claims and those from asbestos.

Motor accidents

For many years when an employee was injured in a motor accident in circumstances where his employer might be liable, case law indicated whether the motor insurer or the employers' liability insurer would handle the claim, but more recently statute has dictated the position. Currently, from 1 July 1994, the Employers' Liability (Compulsory Insurance) Exemption (Amendment) Regulations 1992 SI 1992 No 3172 govern the position. The effect is that an employer's liability for death or bodily injury to an employee is, in the above circumstances, no longer required to be insured under his employer's liability policy. This passes the whole burden of this form of loss to the employer's motor insurers. In the event of there being no motor policy in force, the claim will then be dealt with by the Motor Insurers Bureau as in the case of any other uninsured motor liability.

Damage to employees' property

The employers' liability policy does not cover damage to employees' property, since it is limited to bodily injury to employees only. Thus, damage to an employee's clothing is dealt with by the public liability policy, although as a small item of damage in a personal injury claim it might be paid under the employers' liability policy purely as a matter of administrative convenience in spite of the fact that there is no cover for this item in that particular policy.

Separate clauses defining extra cover

The following clauses are set out separately in the employers' liability policy, and in the case of the contractors' combined policy detailed in Appendix 2 these clauses apply to the public liability Section of that policy as well as to the employers' liability Section. Each of these clauses will be treated in the same way as the clauses of the JCT forms are treated in this book, the clause first being printed in full, with a commentary to follow.

Costs of the insured

(a) Costs
The Company will in addition to the indemnity granted by each section pay
(i)   for all costs and expenses recoverable by any claimant from the Insured
(ii) the solicitors fees incurred with the written consent of the Company for representation of the insured at
   (a) any coroner's inquest or fatal accident inquiry
   (b) proceedings in any Court arising out of any alleged breach of a statutory duty resulting in Bodily Injury or Damage to Property
(iii) all costs and expenses incurred with the written consent of the Company in respect of a claim against the Insured to which the indemnity expressed in this Policy applies.

It is essential for a liability policy to cover any claimant's costs and expenses which can on occasions be as large as the damages awarded.

This sub-paragraph (ii) referring to the solicitor's fees for representation in any court refers to criminal proceedings arising out of breaches of the Factories Acts or similar legislation. The indemnity under the policy only applies to a solicitor's fee incurred with the written consent of the insurer, not to any fine or other penalty imposed by the court.

A successful criminal prosecution in respect of a breach of statutory duty may be given in evidence in a civil action by an injured employee. Thus if an employee is injured by failure to fence a machine in accordance with section 14 of the Factories Act 1961 and the employer is successfully prosecuted by the Inspector of Factories, this conviction may be given in evidence when the employee brings his action in the civil court for damages. Consequently, the insurer and insured have a common interest in any such criminal conviction;

hence the insurer's interest in ensuring that the insured employer is properly defended at the criminal proceedings.

This clause applies to fatal injuries and coroner's court proceedings in a similar way.

All costs and expenses in sub-paragraph (iii) include, for example, examination of defective machinery by engineers after an accident, or expenses of expert witnesses. These, it will be noted, must be incurred with the written consent of the insurer.

Defence costs of individuals

(b) Legal Defence
Irrespective of whether any person has sustained Bodily Injury the Company will at the request of the Insured also pay the costs and the expenses incurred in defending any director manager partner or Employee of the Insured in the event of such a person being prosecuted for an offence under the Health and Safety at Work etc. Act 1974 or the Health and Safety at Work (Northern Ireland) Order 1978.

The Company will also pay the costs incurred with its written consent in appealing against any judgement given.

Provided that
(a) The offence was committed during the Period of Insurance
(b) the indemnity granted hereunder does not
    (i)   provide for the payment of fines or penalties
    (ii) apply to prosecutions which arise out of any activity or risk excluded from this Policy
    (iii) apply to prosecutions consequent upon any deliberate act or omission
    (iv) apply to prosecutions which relate to the health safety or welfare of any Employee unless Section 1 is operative at the time when the offence was committed.
    (v) apply to prosecutions which relate to the health safety or welfare of any person not being an Employee unless Section 2 is operative at the time when the offence was committed.
(c) the director manager partner or Employee shall be subject to the terms exceptions and conditions of the Policy in so far as they can apply.

The 1974 Act codifies many of the rules relating to employers' obligations towards employees in the areas of health, safety and welfare, and makes them the subject of criminal law and not civil law. Failure to comply with the main provisions of the Act will thus lay an employer open to fines or, in serious cases, to imprisonment for a period of up to two years. However, until regulations at present issued under such statutes as the Factories Acts are reproduced under the 1974 Act, they will carry their present civil liability (e.g. the Construction Regulations).

Section 2 of the Act provides that it is the duty of every employer to ensure, so far as is reasonably practicable, the health, safety and welfare at work of all his employees. In detail, the duties are particularised in sub-section (2) as follows:

'(a) the provision and maintenance of plant and systems of work that are, so far as is reasonably practicable, safe and without risk to health;

(b) arrangements for ensuring, so far as is reasonably practicable, safety and absence of risk to health in connection with the use, handling, storage and transport of articles and substances;

(c) provision of such information, instruction, training, and supervision as is necessary to ensure, so far as is reasonably practicable, the health and safety at work of his employees;

(d) so far as is reasonably practicable as regards any place of work under the employer's control, the maintenance of it in a condition that is safe and without risks to health and the provision and maintenance of means of access to and egress from it that are safe and without such risks;

(e) the provision and maintenance of a working environment for his employees that is, so far as is reasonably practicable, safe, without risks to health, and adequate as regards facilities and arrangements for their welfare at work.'

Most insurers have added an extension to their policies to pay the costs of those senior employees who may otherwise have to pay their own costs for defending a prosecution under the Health and Safety at Work etc. Act 1974. In case it is thought that this cover is unnecessary as the senior official's firm would pay for his defence if prosecuted under this Act, it should be noted that this is incorrect, as the Companies' legislation forbids payment of the costs by the accused's own organisation unless the defence is successful. Section 310 of the Companies Act 1985 is the relevant section here, but section 137 of the Companies Act 1989 imports into section 310(3) a new paragraph which effectively legitimises insurance against directors' liability which was in doubt in this respect.

Most prosecutions would arise out of injury to employees and it is usual, therefore, to give this cover under the employers' liability policy. However, the prosecution could arise out of any breach of any sections 2 to 8 of the 1974 Act, without injury to an employee. If Section 1 of the policy is operative there is cover. Therefore this legal costs and expenses cover goes beyond the cover of the employers' liability policy, which is strictly based on a claim arising from bodily injury or disease.

Indemnifying other persons

(c) Indemnity to Other Persons
The Company will indemnify the following as if a separate Policy had been issued to each

(a) in the event of the death of the Insured the personal representatives of the Insured in respect of liability incurred by the Insured

(b) at the request of the Insured
  (i) any officer or member of the Insured's canteen clubs sports athletic social or welfare organisations and first aid fire security and ambulance services in his respective capacity as such
  (ii) any director partner or Employee of the Insured while acting in connection with the Business in respect of liability for which the Insured would be entitled to indemnity under this Policy if the claim for which indemnity is being sought had been made against the Insured.

Provided that
(a) any persons specified above shall as though they were the Insured be subject to the terms exceptions and conditions of this Policy in so far as they can apply
(b) nothing in this extension shall increase the liability of the Company to pay any amount exceeding the Limit of Indemnity of the operative Section(s) regardless of the number of persons claiming to be indemnified.

Paragraph (a) deals with the automatic continuation of policy cover in the event of the death of an insured who is an individual. Without such cover the policy would be inoperative since it is a personal contract between the insured and the insurers. It should be borne in mind that a personal action at law does not cease with the death of either party to it. The Law Reform (Miscellaneous Provisions) Act 1934 provides that if either party to an action dies, the action survives for the benefit of or against the deceased's estate as the case may be.

Paragraph (b) is self-explanatory and is possibly more appropriate to the public liability section of the combined policy considered in the second half of this chapter because it is not usual for claims to be made directly against employees where an employer could be vicariously responsible. Nevertheless, the clause does indicate that if there are circumstances where an employee might be sued, the policy will provide an indemnity.

Provisos (a) and (b) are self explanatory.

Indemnity to principal

(d) Indemnity to Principal
Where any contract or agreement entered into by the Insured for the performance of work so requires the Company will
(a) indemnify the Principal in like manner to the Insured in respect of the principal's liability arising from the performance of the work by the Insured
(b) note the interest of the Principal in the Property Insured by Section 3 to the extent that the contract or agreement requires such interest to be noted.

This extension is a contractual liability clause applying to policy sections 1, 2 and 3. 'Contractual Liability' is defined in the policy in Appendix 2 as follows:

Contractual Liability shall mean liability which attaches by virtue of a contract or agreement but which would not have attached in the absence of such contract or agreement.

A building contractor cannot carry on his business without entering into written contracts. Therefore it is necessary to modify the normal contractual liability exclusion in the policy by extensions of some kind. The cover given by a contractors' policy can be divided into two classes (a) to indemnify the insured in respect of liability he has assumed under the contract and (b) to indemnify the employer directly as a joint insured.

The indemnity in (a) appears in the wording of general exception 2 of the policy in Appendix 2. This is a negative way of giving assumed liability cover by explaining the circumstances where the cover is not given. It reads as follows:

> The company shall not indemnify the Insured under Sections 1 or 2 in respect of Contractual Liability unless the sole conduct and control of claims is vested in the Company but the Company will not in any event indemnify the Insured in respect of
> (i)   liquidated damages or liability under any penalty clause
> (ii) Damage to Property which comprises the Contract Works and occurs after the date referred to in Exception 3 of Section 2 if liability attaches solely by reason of the contract
> (iii) Damage against which the Insured is required to effect insurance under the terms of Clause 21.2.1 of the Joint Contracts Tribunal Standard Form of Building Contract (or any subsequent revision or substitution thereof) or under the terms of any other contract requiring insurance of like kind.

This exception is more applicable to Section 2 of the policy (public liability) than to Section 1 (employers' liability), as it mainly refers to damage to property. Otherwise the exception is self-explanatory. It applies to the obligation on the building contractor under clauses 20 and 21 of the JCT contract.

To return to the second type of contractual liability cover mentioned above, the indemnity in (b), i.e. direct indemnity to the employer, this is given in extension (d) of the policy stated in Appendix 2 and quoted above. Here 'Principal' means the employer who commissions the work in a construction contract. By making the principal an insured party with the contractor the policy complies with the contract which requires the principal to have a direct right of indemnity against the insurer. This is seen in the obligation under clause 22 of the JCT contract.

Exclusions

The modern employers' liability policy comes nearer than any other policy considered in this book to justifying the description 'exclusionless'. The policy does not contain a war risk exclusion, and it is difficult to visualise circumstances in which a legal liability could attach to an employer for injuries arising out of war risks. Sometimes this policy excludes certain hazardous work, e.g. tunnelling and dam or bridge building, but this is work not usually required to be covered by the JCT contract as it is the province of the ICE conditions.

General Exception 1 – Nuclear risks

The Company shall not indemnify the Insured

1.(i) for loss destruction of or damage to any property whatsoever or any loss or expense whatsoever resulting or arising therefrom or any consequential loss

(ii) for any legal liability of whatsoever nature

directly or indirectly caused by or contributed to by or arising from

(a) ionising radiations or contamination by radioactivity from any nuclear fuel or from any nuclear waste from the combustion of nuclear fuel

(b) the radioactive toxic explosive or other hazardous properties of any explosive nuclear assembly or nuclear component thereof.

In respect of Bodily Injury caused to an Employee this Exception shall apply only when the Insured under a contract or agreement has undertaken to indemnify a Principal or has assumed liability under contract for such Bodily Injury and which liability would not have attached in the absence of such contract or agreement.

This is an insurance market clause which appears on all policies because the risks are so great as not to be suitable for ordinary commercial insurance. The Nuclear Installations Acts of 1965 and 1969, together with the Energy Act 1983, make the operator of a nuclear installation solely liable for injury to any person or damage to any third party property. Only the operator is liable, notwithstanding that other parties such as contractors may be liable in tort for the consequences of a nuclear accident on the licensed site. Because the operator alone is liable, he only is required to provide insurance, and the UK Pool policy meets the operator's nuclear legal requirements. This policy is a type of liability policy written through the British Insurance (Atomic Energy) Committee and, inter alia, covers the operator's nuclear liabilities under the statutes mentioned for an overall indemnity figure of £20 million (in relevant cases £5 million) for the current 'cover period'. As from the 1 April 1994 these figures have been increased to £140 million and £20 million respectively for larger and smaller installations. Above £140 million the Government is responsible. The cover period in practice means the lifetime of a licensed installation subject to certain information. This exception applies to an employers' liability policy only when it indemnifies a principal. Such an indemnity undertaken by the contractor means that the principal is being given public liability cover for injury to the contractor's employees, hence the application of this exception. The insured contractor's liability to his employees is covered.

General Exception 2 – Contractual liability

This general exception has already been considered when dealing above with the 'Indemnity to Principal' extension of the policy in Appendix 2.

General Exception 3 – War risks

As previously explained this exception does not apply to the employers' liability policy.

Policy conditions

The conditions will be taken in the order set down in the policy in Appendix 2. This heading has an introductory paragraph which reads:

> This policy and the Schedule shall be read together and any word or expression to which a specific meaning has been attached in any part of this Policy or of the Schedule shall bear such meaning wherever it may appear.

Condition 1 – Alteration in risk

> The Company shall not be liable under this Policy if the risk be materially increased without the written consent of the Company.

The duty of disclosure comes to an end when the policy is issued and does not apply again until renewal since this is, in effect, the negotiation of a new contract.

In order to protect itself against a substantial increase in the risk during the policy period the insurance company uses the above condition. From the wording it is clear that the company is not on cover in respect of any such material change or risk until it has agreed to accept the change. Insurers normally regard as material a basic alteration in design, or an increase in value which considerably exceeds the original estimated contract price, or the amount already provided for in any automatic provision for increasing the sum insured.

Condition 2 – Premium adjustment

> If the premium for this Policy is based on estimates an accurate record containing all particulars relative thereto shall be kept by the Insured.

> The Insured shall at all times allow the Company to inspect such records and shall supply such particulars and information as the Company may require within one month from the expiry of each Period of Insurance and the premium shall thereupon be adjusted by the Company (subject to the Minimum Premium chargeable for the risk being retained by the Company).

This condition sets out the requirements concerning returns the insurance company requires in order to estimate its premiums which are adjustable at the year end when the correct figure is available. The second paragraph allows the company to examine the insured's records concerning wages or turnover on which the premium is based, and reinforces the actions required by the insured in the context of the duty of disclosure within the doctrine of the utmost good faith.

Condition 3 – Duties of the insured

The Insured shall take all reasonable care
(a) to prevent any event which may give rise to a claim under this Policy
(b) to maintain the premises, plant and everything used in the Business in proper repair
(c) in the selection and supervision of Employees
(d) to comply with all statutory and other obligations and regulations imposed by any authority.

The insurers are entitled to expect their insureds to behave with the same care as they would if they were uninsured. However, as the insureds can reasonably expect to be covered when they have failed to take some precautions on the grounds that these circumstances give rise to the very situation for which they arranged insurance, the insurers cannot take too literal an interpretation.

Insurers should not attempt to apply this condition when the insured's employees are negligent as this condition is imposed upon the insured, i.e. the proprietor, whether an individual or a board of directors. Failure to take reasonable care has to be a deliberate, wilful or blatant action on the part of the insured before the insurer can be certain that any attempt to operate the condition will be upheld by a court.

Condition 4 – Make good defects

The Insured shall make good or remedy any defect or danger which becomes apparent and take such additional precautions as circumstances may require.

This condition is self-explanatory and follows from (a) and (b) of the previous condition. For example it may become necessary to insert longer piles if it becomes obvious that the piles inserted or suggested are inadequate.

Condition 5 – Maximum payments

The Company may at any time at its sole discretion pay to the Insured the Limit of Indemnity (less any sum or sums already paid in respect or in lieu of damages) or any lesser sum for which the claim or claims against the Insured can be settled and the Company shall not be under any further liability in respect of such claim or claims except for costs and expenses incurred prior to such payment.

Provided that in the event of a claim or series of claims resulting in the liability of the Insured to pay a sum in excess of the Limit of Indemnity the Company's liability for costs and expenses shall not exceed an amount being in the same proportion as the Company's payment to the Insured bears to the total payment made by or on behalf of the Insured in settlement of the claim or claims.

This condition is self-explanatory, but it should be noted that the proviso cannot apply to the employers' liability policy as there is no limit of indemnity applicable to that cover at the time of writing. However, at renewal in 1995 most insurers offered a limit of £10 million for any one occurrence.

Condition 6 – Claims

The insured or his legal personal representatives shall give notice in writing to the Company as soon as possible after any event which may give rise to liability under this Policy with full particulars of such event. Every claim notice letter writ or process or other document served on the Insured shall be forwarded to the Company immediately on receipt. Notice in writing shall also be given immediately to the Company by the Insured of impending prosecution inquest or fatal inquiry in connection with any such event. No admission offer promise payment or indemnity shall be made or given by or on behalf of the Insured without the written consent of the Company. In the event of Damage by theft or malicious act the Insured shall also give immediate notice to the police.

Condition 6 requires the insured to give notice of any occurrence which may give rise to a claim. This wording, to some extent, leaves the decision to the insured as to whether a claim may arise, but the intention of the insurer is that if there is any doubt at all the claim should be reported to the insurer. In *Kier Construction* v. *Royal Insurance (UK)* (1994) the words 'as soon as possible' were breached by four weeks delay in notification.

The remainder of the condition leaves no decision to the insured. He must send to the insurer immediately every letter, claim, writ, etc. and immediately notify the insurer of any impending prosecution, inquest or fatal inquiry. The notification to the police only concerns Section 3 of the combined policy.

The condition incorporates the usual wording by which insurers can take over the conduct and control of the claim, and prosecute or defend it in the name of the insured. No admission, offer, promise or payment may be made without the consent of the insurer.

Condition 7 – Subrogation

The Company shall be entitled if it so desires to take over and conduct in the name of the Insured the defence or settlement of any claim or to prosecute in the name of the Insured for its own benefit any claim for indemnity or damages or otherwise and shall have full discretion in the conduct of any proceedings and in the settlement of any claim and the Insured shall give all such information and assistance as the Company may require.

This condition increases the insurers' common law rights of subrogation (whereby an insurer can stand in the place of the insured and claim in his or her name after indemnifying him or her) by giving such rights before indemnifying the insured. It also gives the insurer the control of proceedings and settlement of any claim and reminds the insured of his or her duty to give all information and assistance that the company may require.

After the case of *Lister* v. *Romford Ice & Cold Storage Co Ltd* (1957), a market agreement was entered into by all insurers whereby they undertook to take no action to enforce subrogation rights against a negligent fellow employee unless the evidence indicated collusion or wilful misconduct of the employee.

Clause 22.3 of the JCT contract requires the insurers of the works under joint names policies to insure nominated sub-contractors as insured or to waive any right of subrogation which they may have against any such sub-contractors. Similarly in the case of domestic sub-contractors except where the works are existing structures.

Condition 8 – Contribution

> If at the time of any event to which this Policy applies there is or but for the existence of this Policy there would be any other insurance covering the same liability or Damage the Company shall not be liable under this Policy except in respect of any excess beyond the amount which would be payable under such other insurance had this Policy not been effected.

This wording adopts the more modern approach of not merely stating that, if there is another insurance in operation covering the same subject matter against the same risk in respect of the same insured, it will not apply, as in addition it emphases that it will only pay any excess beyond the amount which would be payable under such other insurance. Without this emphasis the effect is the same as if the clause had said it would only pay its rateable proportion assuming both insurers say they are not going to contribute and both their policies are in operation. The authority for this is *Gale* v. *Motor Union Insurance Co* and *Loyst* v. *General Accident Fire and Life Assurance Corporation Ltd* (1928) (two cases heard together). The courts will not allow the two non-contribution policy conditions (which are exactly the same) to cancel the cover given with the result that if the loss is covered elsewhere it is covered nowhere.

The correct decision must be that each policy should contribute its rateable proportion. However, once the wording is different so that one policy contains, as here, an excess wording in the contribution condition stating that it will not operate until the second policy is exhausted and the second policy does not contain such a clause, then the common law principle of contribution will not operate between the two policies (see the New Zealand Court of Appeal decision in *State Fire Insurance General Manager* v. *Liverpool London and Globe Insurance Co Ltd* (1952).

On the other hand, from the Scottish appeal case of *Steelclad Ltd* v. *Iron Trades Mutual Insurance Co Ltd* (1984) where *both* policies contained an excess wording it is clear that both will have to contribute on a rateable proportion basis. For further details of this case see Chapter 2 under the heading 'Insurance law'.

An illustration of the basic rule that for contribution to apply the same risk and the same property must be covered for the same interest appears in the damage' are qualified by the word 'accident' or 'accidental' in the operative clause. Alternatively, if the word 'accident' or 'accidental' does not appear in the operative clause, it is almost certain that an exception concerning inevitable or deliberate injury, loss or damage will take its place.

Consequential loss arising from this injury or damage is covered.

A commentary on this policy is difficult because insurers vary considerably in the subject matter of the exclusions and in the actual wording when the meaning is similar.

Accident

Accident has been legally defined in *Fenton* v. *Thorley* (1903) as 'an unlooked-for mishap or an untoward event which is not expected or designed'. Thus, the phrase 'accidental injury or damage' includes injury or damage happening by chance, unexpectedly or without design, and is not dependent upon one specific event, whereas the phrase 'injury or damage caused by accident', used by some insurers, is considered to be limited in this way.

By 'bodily injury' most insurers intend to include death, disease and illness.

The significance of the words 'accidental' or 'accident' appearing in the operative clause on the one hand, and an exception concerning inevitable or deliberate injury or damage on the other lies in the fact that in the former case the onus of proving an accident or an accidental occurrence is upon the insured, whereas in the latter case it rests upon the insurer to prove the exception applies. The authorities for this statement are *Munro Brice & Co* v. *War Risks Association* (1918), which decided that the burden of proof is upon the insured to prove an accident when these words appear in the operative clause of the policy, and the case of *Bond Air Services* v. *Hill* (1995) decided that the onus is upon the insurer to prove the operation of an exclusion. The words 'exclusion' and 'exception' are synonymous.

Consequently, the exception as described provides slightly wider cover than the operative clause with the accidental qualification on those few occasions when the circumstances are equally consistent with accident or no accident, because the party upon whom the onus of proof falls will fail since it is impossible to discharge the onus (see *Wakelin* v. *London Western Railway* (1886)).

More important than the point just made is the interpretation given to the words 'inevitable' or 'deliberate' damage when these are used in an exclusion. However, it is probably sufficient to say that the word 'accidental' has been legally defined as already mentioned, whereas the exclusions of 'inevitable' and 'deliberate' damage have no legal definition and the insured is to some extent in the insurer's hands concerning interpretation.

It should be appreciated that the phrase 'accidental loss of or damage to property' could include deliberate theft by the contractor's employee of the building owner's property, as arguably it is accidental from the insured contractor's viewpoint unless, for example, he knew that the guilty employee had a criminal record and that he had carried out similar acts while in the insured's employment when the insured might be guilty of lack of reasonable precautions (a policy condition). Some insurers make the point clear by a specific exclusion in the policy of loss or damage due to theft with a proviso covering theft caused by the negligence of the insured or by an employee of the insured. Such a proviso covers the circumstances in the cases of *Stansbie* v. *Troman* (1948) and *Petrovitch* v. *Callinghams* (1969) mentioned in Chapter 1 under the heading of 'Negligence'.

Accidental obstruction, trespass or nuisance

Claims for loss of business due to obstruction of access to premises or loss of production caused by failure of the electricity supply due to the insured contractors' negligent actions are fairly frequent. Three types of claims can be considered from:

(1) the electricity, water or gas authorities for damage to their cables or pipes;
(2) occupiers of shops or garages for loss of business due to obstruction of the highway, although they suffer no physical damage;
(3) a nearby factory owner who has lost production because of the cut in the power supply.

Assuming the contractor is negligent and that the construction contract concerned does not affect the legal position then the policy position in regard to each of these three cases is as follows:

(1) Damage to cables, etc, is damage to property within the meaning of (b) in the operative clause.

(2) Occupiers of premises adjoining the highway may recover in an action based on public nuisance when they suffer special damage as a result of a nuisance on the highway over and above the inconvenience suffered by the users of the highway. In *Fritz* v. *Hobson* (1880) an unreasonable obstruction of a private way to the home and shop of an antique dealer was held to be actionable because it resulted in loss of custom to him. While (a) and (b) of the operative clause would not cover such a claim because there was no bodily injury or damage to property, (c) would cover such incidents.

(3) Pure economic loss claimed in negligence is not recoverable in current law unless it flows directly from injury or damage to property (see *Spartan Steel & Alloys Ltd* v. *Martin & Co (Contractors) Ltd* (1972)). The exceptions to this rule are the decisions in the cases of *Hedley Byrne & Co Ltd* v. *Heller & Partners Ltd* (1982) (when there can be liability for pure economic loss arising from negligent mis-statements), and *Junior Books Ltd* v. *Veitchi Co Ltd* (1982) (which was decided mainly on the relationship between the two parties to the action being employer and nominated sub-contractor, which although

not contractual was very close to it). The point in the second case arose from the fact that there can be liability for pure economic loss where a contractual relationship exists between the two parties. However, neither of these exceptions apply to (3) above so there would be no legal liability unless the loss of production flowed directly from damage to the claimant's property, when the policy would operate under (b) of the operative clause. Nevertheless attention is drawn to the expansion of the *Hedley Byrne* position as explained in Chapter 1 by virtue of the two construction cases of *Conway* v. *Crowe Kelsey* (1994) and *Barclays Bank plc* v. *Fairclough Building Ltd* (1994). However, it is not thought that these decisions would alter the remarks above concerning the position in (3) above as there is no special relationship of proximity between the factory owner and the contractor employed by another party who is working in a road some distance away.

See Chapter 1 for an explanation of trespass and nuisance.

Territorial limits

This term has already been explained earlier in this chapter.

'The Insured shall be legally liable to pay'

Strictly speaking this means that the insurer is under no obligation to pay a claim by a third party until liability of the insured has been established by the courts. See *Post Office* v. *Norwich Union Fire Insurance Society Ltd* (1967), which concerned a third party claiming under the Third Party (Rights Against Insurers) Act 1930, against the insurers of an allegedly negligent contractor, who had gone into liquidation. In practice insurers take over the handling of a third party claim immediately they are informed of the occurrence as they wish to control the situation and any costs involved. The majority of claims are handled without litigation.

As already mentioned 'legal liability' and 'liability at law' (used by some insurers) are synonymous; in *MS Aswan Engineering Establishment Co* v. *Iron Trades Mutual Insurance Co Ltd* (1988), the Iron Trades argued that it did not include contractual liability but the court decided otherwise. The judge said that the meaning of 'liability at law' was to be determined by reference to the ordinary use of language. It should not be given any restricted meaning to accord with the insurer's intentions if the words used in the insurers standard form give rise to any doubt. Thus the term was not restricted to liability in tort. See Chapter 1 for further details of legal liabilities.

The limit of indemnity in the public liability policy is specified in the policy schedule in Appendix 2. The amount payable in most public liability policies in any one period of insurance is usually unlimited, but a limit is applied to any one occurrence or series of occurrences arising out of one event. As a generalisation there is no limit to the amount of damages a court might award.

Additional covers

In Appendix 2 in Section 2 (the public liability section) the following additional covers appear after the operative clause.

Motor contingent liability

Notwithstanding Exception 2(c) below the Company will indemnify the Insured within the terms of this Section in respect of liability for Bodily Injury or Damage to Property caused by or through or in connection with any motor vehicle or trailer attached thereto (not belonging to or provided by the Insured) being used in the course of the Business

Provided that the Company shall not be liable for
(a) damage to any such vehicle or trailer
(b) any claim arising whilst the vehicle or trailer is
    (i)   engaged in racing pacemaking reliability trials or speed testing
    (ii) being driven by the Insured
    (iii) being driven with the general consent of the Insured or of his representative by any person who to the knowledge of the Insured or other such representative does not hold a licence to drive such a vehicle unless such a person has held and is not disqualified from holding or obtaining such a licence
    (iv) used elsewhere than in Great Britain Northern Ireland the Isle of Man or the Channel Islands.

Contractors may incur liability, known as contingent liability, in respect of a vehicle which is being used on the contractor's behalf but over which he has no direct control. For example, the contractor who permits his employee to use the employee's own car on his (the contractor's) business or the contractor who hired from the owner a vehicle with a driver.

The Road Traffic Act requires that the person who has effective control of the vehicle, i.e. the person who uses the vehicle or causes or permits any other person to use the vehicle, shall effect third party insurance. This means in the above examples that the employee and the owner of the hired out vehicle (with driver) are the people who should take out insurance. While the policies issued to these persons can be extended to indemnify the contractor, there are dangers that this cover may not operate to the contractor's advantage. For example, such policies may lapse, they may be invalid, or they may not cover the particular use which the contractor requires. Therefore, a contingent liability cover is necessary, however unlikely it may be considered that such a contingency might arise.

The wording obviously has to override exception 2(c) (see below) and the provisos are self-explanatory.

Defective Premises Act 1972

This clause reads:

The indemnity provided by this Section shall extend to include liability arising under Section 3 of the Defective Premises Act 1972 or Section 5 of the Defective Premises (Northern Ireland) Order 1975 in respect of the disposal of any premises which were occupied or owned by the Insured in connection with the Business

Provided that the Company shall not be liable for the cost of remedying any defect or alleged defect in such premises.

Under the Defective Premises Act liabilities might arise out of premises which the insured contractor has disposed of and which are not specified in the policy schedule.

The introductory paragraph of section 3 of this Act reads as follows:

Where work of construction, repair, maintenance or demolition or any other work is done on or in relation to premises, any duty of care owed, because of the doing of the work, to persons who might reasonably be expected to be affected by defects in the state of the premises created by the doing of the work shall not be abated by the subsequent disposal of the premises by the person who owed the duty.

The above quoted clause in the policy is intended to deal with this potential liability in respect of property no longer owned or occupied by the insured. This cover applies to accidents occurring during the policy period even though the defects may have arisen earlier.

Movement of obstructing vehicles

Exception 2(c) shall not apply to liability arising from any vehicle (not owned or hired by or lent to the Insured) being driven by the Insured or by any Employee with the Insured's permission whilst such vehicle is being moved for the purpose of allowing free movement of any vehicle owned hired by or lent to the Insured or any Employee of the Insured

Provided that
(a) movements are limited to vehicles parked on or obstructing the Insured's own premises or at any site at which the Insured are working
(b) the vehicle causing obstruction will not be driven by any person unless such person is competent to drive the vehicle
(c) the vehicle causing obstruction is driven by use of the owner's ignition key
(d) the Company shall not indemnify the Insured against
  (i) damage to such vehicle
  (ii) liability for which compulsory insurance or security is required under any legislation governing the use of the vehicle.

This clause is self explanatory although not standard in the insurance world. Incidentally exception 2(c) excludes liability arising out of the ownership possession or use by the insured of mechanically propelled vehicles licensed for road use (subject to certain exceptions) known as Road Traffic Act liability.

Leased or rented premises

This clause reads

> Exception 4(b) shall not apply to damage to premises leased or rented to the Insured
>
> **Provided that** the Company shall not indemnify the Insured against
> (a) Contractual Liability
> (b) the first £100 of Damage caused otherwise than by fire or explosion

One effect of the custody or control exception (exception 4(b) in the policy under discussion) is that if the insured is the tenant of property then his legal liability for damage to that property is excluded. Now the tenant may be liable because of his negligence or contractually under the terms of the lease. The effect of the clause quoted above is to cover the insured's liability for negligence for such damage, but not contractually under the lease, as contractual liability is defined under the 'Definitions' section of the policy as follows:

> **Contractual Liability** shall mean liability which attaches by virtue of a contract or agreement but which would not have attached in the absence of such contract or agreement.

If the Insured requires this cover it can be done by taking out a separate fire policy in his own name or by arranging such a policy in the joint names of the landlord and tenant, which will prevent the fire insurers from exercising subrogation rights against the tenant.

Proviso (b) is self-explanatory.

Exceptions

Each of these exceptions is prefaced by the words

> The Company shall not indemnify the Insured against liability ...

Exception 1 – Liability to employees

> in respect of Bodily Injury to any Employee arising out of and in the course of his employment by the Insured.

The obvious reason for this exception is that the employers' liability policy covers this liability.

Exception 2 – Risks insured more specifically by other policies

> arising out of the ownership possession or use by or on behalf of the Insured of any
> (a) aircraft aerospatial device or hovercraft
> (b) watercraft other than hand propelled watercraft or other watercraft not exceeding 20ft in length
> (c) mechanically propelled vehicle licensed for road use including trailer attached thereto other than liability caused by or arising out of

(i)   the use of plant as a tool of trade on site or at the premises of the Insured
(ii) the loading or unloading of such vehicle
(iii) damage to any building bridge weighbridge road or to anything beneath caused by vibration or by the weight of such vehicle or its load

but this indemnity shall not apply if in respect of such liability compulsory insurance or security is required under any legislation governing the use of vehicle.

Aircraft or watercraft are excluded because they should be insured in the aviation or marine market: similarly with any mechanically propelled vehicle used in circumstances to which the Road Traffic Acts apply. However, a public liability policy may be extended to cover what is known as the 'tool of trade' risk attaching to plant where this risk is not covered by the motor policy. Loading and unloading risks not covered by the motor policy should also be included in the public liability cover. The third exception made to the mechanically propelled vehicle exclusion is to cover an exception which used to apply universally (but not nowadays) to the commercial vehicle policy concerning liability for damage to buildings bridges and weigh bridges etc. Nevertheless, the obligatory insurance required by the Road Traffic Acts is still a matter for the motor policy.

Exception 3 – Contract works

for Damage to Property which comprises the Contract Works in respect of any contract entered into by the Insured and occurring before practical completion or a certificate of completion has been issued.

'Contract Works' is defined in the definition section of the policy as follows:

Contract Works means the temporary or permanent works executed or in course of execution by or on behalf of the Insured in the development of any building or site or the performance of any contract including materials supplied by reason of the contract and other materials for use in connection therewith.

Contract Works as defined is properly insured under the contractors' all risks policy.

Exception 4 – Property owned by or in the custody or control of the insured

in respect of Damage to Property
(a) belonging to the Insured
(b) in the custody or under the control of the Insured or any Employee (other than Property belonging to visitors directors partners or Employees of the Insured)

Exception 4(b) shall not apply to Damage to buildings (including contents therein) which are not owned or leased or rented by the Insured but are temporarily occupied by the Insured for the purpose of maintenance alteration extension installation or repair.

Liability for damage to property belonging to the insured should be covered by material damage policies as should property in the insured's custody or control. For contractors working on the premises of third parties the exception is qualified to make it clear that such premises, and their contents are for the purposes of the public liability policy not to be considered as in the custody or control of the insured.

Exception 5 – Defective workmanship, etc.

> for the cost of and expenses incurred in replacing or making good faulty defective or incorrect
> (a) workmanship
> (b) design or specification
> (c) materials goods or other property supplied installed or erected
> by or on behalf of the Insured but this Exception shall not apply to accidental Damage which occurs as a direct consequence to the remainder of the Contract Works which are free of such fault defect or error.

While liability insurers do not intend to pay for the replacement of defective workmanship, and (b) and (c) are akin to defective workmanship, the consequences of such defective work are covered. It is difficult to see the reason for the reference to contract works in the final paragraph as such works are excluded by Exception 3 mentioned earlier.

In the case of *Lester* v. *White* (1992) a contractor's third party liability policy in New Zealand excluded liability for 'defect error or omission in design, plan, specification or formula'. Foundations failed due to failure to make provision for reinforcement in the floor, but there was no actual specification or detailed design of the floor slab, which had proved unsuitable for a house built on filled ground. In the absence of some such document it was argued that the exclusion did not apply. The judge held that had a claim in tort against the contractors (which in fact failed for discoverability and limitation reasons) succeeded, it would have been successfully excluded from the risks insured.

Exception 6 – Breach of professional advice given for a fee

> caused by or arising from advice design or specification provided by or on behalf of the Insured for a fee.

This is particularly necessary where a firm of contractors have their own architects department. These risks are covered by the professional indemnity insurance market. Probably most policies would not include the words 'given for a fee' as in almost all cases the advice would inevitably be given as part of the contract price in a design and build contract.

Exception 7 – Excess

> for the Excess specified in the Schedule other than for Damage to premises leased or rented by the Insured.

Contractors' public liability policies sometimes include a large excess in respect of underground services as well as a separate excess (not usually applicable to bodily injury claims) for general third party claims. The excess is not usually applicable to bodily injury claims as insurers wish to have full control of these claims. It does not apply to premises leased or rented by the insured as there is already an excess in operation under this cover.

Exception 8 – Pollution

caused by or arising from seepage pollution or contamination unless due to a sudden unintended and unexpected event.

The effect of this clause is that sudden accidental pollution is covered, otherwise this risk is not covered. It follows that if the House of Lords had found the defendants liable in *Cambridge Water Company* v. *Eastern Counties Leather* (1994), prima facie on the facts of that case they would not have been covered on a policy with this wording. See Chapter 1.

Precautions required in certain circumstances

These precautions are conditions precedent to the liability of the insurers in circumstances which arise out of the activities of contractors and could be, and by some insurers are, put under the conditions section of the policy.

Precaution 1 – Use of heat

This clause reads as follows:

It is a condition precedent to the liability of the Company that when
(a) welding or flame-cutting equipment blow lamps blow torches or hot air guns are used by the Insured or any Employee away from the Insured's premises the Insured shall ensure that
   (i)   all moveable combustible materials are removed from the vicinity of the work
   (ii) suitable portable fire extinguishing apparatus will be kept ready for immediate use as near as practicable to the scene of the work
   (iii) before heat is applied to any wall or partition or to any material built into or passing through a wall or partition an inspection will be made prior to commencement of each period of work to make certain that there are no combustible materials which may be ignited by direct or conducted heat on the other side of the wall or partition
   (iv) they are lit as short a time as possible before use and extinguished immediately after use and that they are not left unattended whilst alight
   (v) blow lamps are filled and gas cylinders or canisters are changed in the open
   (vi) the area in which welding or flame-cutting equipment is used will be screened by the use of blankets or screens of incombustible material
   (vii) a fire safety check is made in the vicinity of the work on completion of each period of work

(b) vessels for the heating of asphalt or bitumen are used away from the Insured's premises the Insured shall ensure that each vessel
  (i) shall be kept in the open whilst heating is taking place
  (ii) shall not be left unattended whilst heating is taking place
  (iii) if used on a roof shall be placed on a surface of non-combustible material
  (iv) shall be suitable for the purpose for which it is intended and be maintained and used strictly in accordance with the manufacturer's instructions.

The size and extent of third party claims from fires caused by negligent contractors are such that the insured and his employees are required to take reasonable precautions in the conduct of their operations to prevent the outbreak of fire. The wordings of this clause are set out in considerable detail and are therefore self-explanatory.

Precaution 2 – Property in the ground

This clause reads

The indemnity provided by this Section shall not apply to liability in respect of Damage to pipes cables mains and other underground services unless the Insured
1. has taken or caused to be taken all reasonable measures to identify the location of pipes cables mains and other underground services before any work is commenced which may involve a risk of Damage thereto
2. has retained a written record of the measures which were taken to comply with 1. above before such work has commenced
3. has adopted or caused to be adopted a method of work which minimises the risk of Damage to such pipe cables mains and other underground services.

Damage to property in the ground is a common cause of claims under this policy. Telephone and electricity cables, gas pipes and water mains are damaged by excavators. Consequently insurers try to impose some care in the approach of contractors to road works and other work which may involve damage to these services.

Extensions

(a) Costs, (b) Legal defence, (c) Indemnity to other persons and (d) Indemnity to Principal

See earlier in this chapter under the heading 'Separate clauses defining extra cover'. However, the next extension only applies to the public liability policy.

(e) Cross liabilities

This clause reads:

The company will indemnify each insured to whom this Policy applies in the same manner and to the same extent as if a separate policy had been issued to each provided that the total amount of compensation payable shall not exceed the Limit of Indemnity regardless of the number of persons claiming to be indemnified.

Provided that the Company shall not indemnify the Insured against liability for which an indemnity is or would be granted under Employers Liability Insurance but for the existence of this Policy.

A public liability policy covering joint insureds requires a cross liabilities clause because it does not cover damage to property owned by the joint insured and often excludes property in his custody or control. Also it does not cover injury to persons under a contract of service of apprenticeship with the employer commissioning the work because he is a joint insured. All this is to the contractor's disadvantage as the property damaged and injuries concerned are claims which should be covered by the public liability policy because the claimants are third parties to the contractor. Similarly the employer who commissions the work in a construction contract has no cover in respect of this legal liability for damage to property belonging to the contractor or in the contractor's custody or control, nor for injury to persons under a contract of service or apprenticeship with the contractor. This results in a reduction of cover to the employer given by the contractor's policy. All these restrictions are overcome by a cross liabilities clause which construes the policy as though separate policies had been issued to each of the joint insureds.

However, it has to be made clear that this does not mean that there are two limits of indemnity as only the one applies. Similarly it is made clear that if an employers' liability policy applies, the cross liabilities clause does not alter that position. Furthermore clause 21 of the JCT contract does not require a joint names policy unlike the ICE conditions' equivalent clause.

General Exceptions

Exception 1 – Nuclear risks and 2 – Contractual liability

See earlier in this chapter when discussing the employers' liability policy.

Exception 3 – War and kindred risks

under Section 2, 3 or 4 for any consequence of war invasion act of foreign enemy hostilities (whether war be declared or not) civil war rebellion revolution insurrection or military or usurped power.

This again is an insurance market clause as it deals with matters that a commercial insurance policy could not underwrite for this type of cover. War risks were handled by the Government in the First and Second World Wars.

So far as kindred risks 'invasion act of foreign enemy hostilities (whether war be declared or not)' are concerned, they all imply the existence of a state of war with a foreign power, and the pattern of international events during and since the last world war is reflected in the qualification 'whether war be declared or not'. The exclusion is worded so that it applies not only to losses due to enemy acts but also to those incurred through the steps taken to oppose the enemy, as shown in the Falklands crisis. Clauses 32 and 33 of JCT 80 deal respectively with the procedure in the event of an outbreak of hostilities and how war damage is to be handled. Civil war implies something in the nature of organised acts of warfare.

Rebellion in *Lindsay and Pirie* v. *General Accident etc. Corporation Ltd* (1914) was defined as the taking up of arms traitorously against the Crown, whether by natural subjects or others when subdued. It can also mean disobedience to the process of the law as applied by the courts. Revolution is very similar.

Insurrection is a stage short of rebellion and is defined in *Jowitt's Dictionary of English Law* as a rising of the people in open resistance against established authority with the object of supplanting it.

The term 'military power' includes acts done by the Crown's military forces in opposition to subjects of the realm in open rebellion and organised as a military force. 'Usurped power' applies to an organised rebellion which is acting under some authority and has assumed the power of government by making laws and enforcing them. In *Curtis* v. *Mathews* (1919), Banks LJ said:

'Usurped power seems to me to mean something more than the action of an unorganised rabble. How much more I am not prepared to define. There must probably be action by some more or less organised body with more or less authoritative leaders.'

It has been suggested that the last four words of the kindred risks to war, namely 'military or usurped power' come within the exclusion of nuclear weapons material used outside the UK.

Conditions of the policy

See earlier in this chapter when considering the employers' liability policy.

6.3 *Clauses 20 and 21 and the liability policies*

Summary of the main risks remaining uninsured

In order to indicate the main risks for which the parties to the JCT 80 contract may not have insurance cover, the positions of the contractor and employer are considered in turn. However, it is emphasised that it is the basic policy cover that has been described in this chapter which is considered here, not the cover that an insurance broker by negotiation with an insurer could produce for one particular insured or tailor-made for one particular contract. Also, there is no attempt to indicate certain special wordings which some brokers catering for the construction industry have persuaded an insurer or insurers to adopt for the former's clients.

The contractor

Limit of indemnity

The contract in sub-clause 21.1.1.2 (part of the insurance clause) will specify a minimum limit for any one occurrence or series of occurrences arising out of one event for liability to third parties (other than employees). However, the contract in clause 20 (the indemnity clause) and the position at common law does not limit liability to third parties. Rarely, if ever, will insurers cover the liability in respect of any one occurrence, or series of occurrences arising out of one event, up to an unlimited amount, although very high limits can be obtained.

The meaning of 'property' in the operative clause of the public liability policy

The word 'property' should be defined or qualified in the policy (as in Appendix 2) and if it is not, the insurer should be asked for his definition. It is foolish to wait for a claim to arise involving a borderline definition of the word 'property' before the meaning of the word is considered. It is better to understand, from the commencement of the insurance contract, the limitation which will no doubt be placed on the meaning of this word. It is also preferable for some cover to be given for accidental obstruction or trespass, as mere 'loss or damage to property' cover may not be interpreted to cover financial loss resulting from such obstruction or trespass.

Cancellation of the policy

Only the insurer has the right of cancellation, not the insured, and the condition must be adhered to literally, i.e. notification must be given by registered letter. Although it is rarely used, it has to be pointed out that the JCT contract does not allow for the very infrequent occasion on which the condition might be invoked. The insurer can be asked to delete the condition, and the very fact that it is rarely needed is an argument for deletion. An insurer might be prepared to fall back on his right to refuse to renew.

The gap between the public liability policy and the contractors' all risks policy concerning the contract works and future structural damage

The public liability policy excludes loss or damage to the contract works, and as will be seen in the next chapter the CAR policy usually lapses when the contract is completed, including any liability under the defects liability period. Yet it is clear that there is a liability for many years in tort and contract after the contract is completed for defective work and materials and for defective design (a CAR policy exclusion). Neither the certificate of practical completion (clause 17.1) nor the certificate of the completion of making good defects (17.4) contains any permanent binding force under JCT 80, but see the cases on the effect of the Final Certificate on pages 240 and 241.

The employer

The employer's responsibility to third parties because he commissioned the works

Under clause 20 the contractor does not have to indemnify the employer for the latter's 'act of neglect' including that of any person for whom the employer is responsible for personal injuries. Similarly (apart from the application of clauses 20.3 and 22C.1), the contractor only has to indemnify the employer to the extent that he, and those for whom clause 20.2 makes him responsible, are liable for negligence, breach of statutory duty, omission or default. Consequently the employer will, under clause 21.1 (the insurance clause), get no direct insurance cover from the contractor's liability policies as they only protect the contractor in respect of the indemnity which the contract so requires, and that is as set out above. Thus the contract only gives the employer insurance protection when he claims under the contract against the contractor in the circumstances set out above, which is an indirect type of limited cover. The employer cannot claim directly against the contractor's insurer under the policies.

The employer therefore has the responsibility of seeing that he has liability insurance cover in respect of any persons employed or engaged directly by him and for his own work (if he does any) in respect of the project. Consequently, unless the employer makes special arrangements for liability cover with his insurers, he is unlikely to be covered by his normal annual liability policies unless he operates in the construction industry, e.g. he is a building developer or local authority. If, for example, he is merely having a factory built and his normal occupation is not construction work, he will only be insured in respect of that occupation, e.g. biscuit manufacture, and special insurance arrangements are necessary. It is appropriate to mention that the employer can obtain cover for subsidence risks to third party property by making the necessary arrangements under sub-clause 21.2.1; details of this cover will be given in later chapters.

Consequential loss

The public liability policy covers any consequential losses flowing from the injury to persons or damage to property covered by the operative clause, although the policy does not specifically say so. The problem is that although there is an exclusion the purpose of which is to exempt loss or damage to the contract works, insurers claim that this means that consequential losses arising from such loss or damage must also be excluded. It really depends how the exclusion is worded. If the wording is that 'liability in respect of loss or damage to the contract works' is excluded, then it is probably also sufficient to exclude the consequential loss. There is an Australian authority that the words 'in respect of' are difficult of definition but have the widest possible meaning of any expression intended to convey some connection or relation between the two subject matters to which the words refer (*Trustees Executors and Agency Co Ltd* v. *Reilly* (1941)).

Cancellation of the policy

The same remarks apply here as were made under this heading when considering the contractor's position.

Chapter 7

Contractors' all risks policy

The contractors' all risks policy is said to have come into existence as a result of the Institution of Civil Engineers' Conditions of Contract. When these conditions first required the contract works to be insured on an all risks basis, insurers began to provide this policy. In spite of its name (All Risks) this policy has its exceptions and limitations. Possibly 'Contract Works Insurance' is a better title, but the JCT contract uses the term 'All Risks Insurance' so it will be used here. This material damage cover called for by clause 22 of JCT 80 as amended, is defined in clause 22.2, except for clause 22C.1 which only requires cover for the specified perils defined in clause 1.3. The latter is known in insurance circles as a fire and special perils policy which is considered in Chapter 12. Clause 22.3 concerning cover for sub-contractors also only refers to specified perils.

It is sufficient to mention at this stage that the contractors' all risks policy, as its name implies gives a wider cover. For that reason, apart from the requirements of the particular contract, it is a better form of protection. Attention is drawn to Appendix 2, Section 3, where a typical contractors' all risks policy (hereafter referred to as the CAR policy) will be found. To avoid repetition those parts of the combined policy which were considered in the previous chapter will not be detailed here.

7.1 The policy clauses

The operative clause

In the event of Damage to the Property Insured the Company will by payment or at its option by repair reinstatement or replacement indemnify the Insured against such Damage

Provided that
1. the Company shall not indemnify the Insured
    (a) under Items 1, 2, 3 and 5 of the Property Insured for any amount exceeding the Limit of Indemnity in respect of each Item in any one Period of Insurance
    (b) under item 4 of the Property Insured for any amount exceeding the Limit of Indemnity in respect of any one item

2. the Property belongs to or is the responsibility of the Insured
3. the Property is
   (a) on or adjacent to the site of the Contract Works within the Territorial Limits or
   (b) in transit within the Territorial Limits by road (whether under its own power or otherwise) rail or inland waterway or
   (c) elsewhere within the Territorial Limits (but not Item 1 of the Property Insured) and stored in a locked premises or compound.

'In the event of Damage to the Property Insured'

Under the heading 'Definitions' damage includes loss and property means material property.

The property insured appears in the schedule of the policy as follows:

Item 1 – Contract Works
Item 2 – Constructing Plant Tools and Equipment owned by the Insured
Item 3 – Temporary Buildings and Site Huts (including fixtures and fittings therein)
Item 4 – Hire-in or borrowed Property described in items 2 and 3 not exceeding £... any one item
Item 5 – Personal Effects and Tools of the Insured's Employees not exceeding £... any one Employee

The policy defines contract works as follows:

Contract Works means the temporary or permanent works executed or in course of execution by or on behalf of the Insured in the development of any building or site or the performance of any contract including materials supplied by reason of the contract and other materials for use in connection therewith.

'The Company will ... indemnify'

The alternative methods of indemnity to a payment, i.e. repair, reinstatement or replacement, will in practice be a payment to the insured for the cost of repairing or replacing the damage to the insured property.

'The Insured'

To comply with most construction contracts the insured will be the employer and the main contractor although other parties such as sub-contractors may be included as the insured in the schedule of the policy. However the prudent contractor will insure his property including the works even when not required to do so by contract conditions. Further, the 'all risks' cover could well be more extensive than the perils required to be insured by contract.

'Against such damage'

Damage is defined as including loss, and all forms of loss or damage are covered subject to the policy exceptions and conditions (see later).

Proviso 1

Clearly the intention here is to limit the maximum amount payable in respect of each item of property to the amount stated in the schedule in each period of insurance which is also stated in the schedule. The object is also to separate hired-in property so emphasising that the limitation is in respect of each item of plant etc. that is hired-in, and is to the amount stated in the schedule.

Proviso 2

Obviously all the property listed in the items 1 to 5 of the schedule either belongs to or is the responsibility of the insured under the construction contract otherwise there would be a lack of insurable interest and the property concerned could not be insured by those named as insured in the policy schedule.

Proviso 3

Unlike proviso 2 it is not so obvious that all the property listed in the items 1 to 5 is either on or adjacent to the site of the contract works or is being carried by road, rail or inland waterway to or from the site of the contract works within the territorial limits. Some materials or goods might be stored off site and it will be seen later that if they have been certified for payment under the contract terms the indemnity provided by the policy is extended to apply to such goods or materials as being intended for incorporation in the contract.

The territorial limits were considered in the previous chapter, but it should be noted that transit by sea or air of the property is not included, which is because of the heavier risks involved which in turn call for specialised underwriting considerations. Secondly there is a requirement that all items in the policy schedule other than item 1 (the Contract works) must be stored in a locked premises or compound, when not on or adjacent to the site or in transit.

Additional covers

Additional cover 1 – Professional fees

The Company will in addition to the Limit of Indemnity indemnify the Insured for architects surveyors consulting engineers and other professional fees necessarily incurred in the repair reinstatement or replacement of Damage to the Property Insured to which the indemnity provided by this Section applies

Provided that
(a) such fees shall not exceed that authorised under the scales of the appropriate professional body or institute regulating such charges
(b) the company shall not indemnify the Insured against any fees incurred by the Insured in preparing or contending any claim.

It is important that the limit of indemnity (sum insured) estimated by the insured for the contract works, including temporary works, should make provision for these fees. Some construction contracts such as clause 22 of the JCT 80 contract and clause 21 of the ICE conditions require this risk to be covered by the contractors' all risks policy.

Proviso (a) is self-explanatory and proviso (b) emphasises the fact that insurers are not going to pay to assist the insured by financing claims against themselves.

The words 'professional fees necessarily incurred in the repair reinstatement or replacement of Damage to the Property Insured' emphasise that the cost of a site investigation will be claimable under the policy if it is related to the repair reinstatement or replacement of damage to the property insured to which the indemnity provided by the policy applies.

Sometimes professional fees are duplicated. Thus a surveyor or architect dealing with the repair of subsidence damage involving the employment of a consulting engineer to advise on, and prepare a scheme for, underpinning, can result in the architect or surveyor submitting the full professional scale fee on all the work, in addition to the engineer's fees, as if the latter had not been appointed.

In the first place if additional work, over and above the repair reinstatement or replacement of the original work, is necessary, the cost of such additional work and fees connected with it are not covered by the policy. The policy does not cover defective or incorrect workmanship, design or specification, or materials or goods installed, erected or intended for incorporation in the contract works. See Exception 1 later in this chapter.

Secondly, even if the work does only involve the repair reinstatement or replacement of existing work and the surveyor or architect does not perform the complete duties specified in the RICS or RIBA scales of professional charges, because the engineer does some of the work, the surveyor or architect is not entitled to the complete percentage scale fee.

The onus of proof is on the insured that the costs relate to the repair, reinstatement or replacement of damage to the property insured as these words come from the operative clause of the policy. They must provide evidence which will mean incurring costs. However, if they employ a claim maker or adviser, such as an assessor, the fees will not be covered (see proviso (b)).

Additional cover 2 – Debris removal

The Limit of Indemnity provided in respect of Item 1 of the Property Insured shall

extend to include the cost and expenses necessarily incurred by the Insured with the consent of the Company in
(a) removing and disposing of debris from or adjacent to the site of the Contract Works
(b) dismantling or demolishing
(c) shoring up or propping
(d) cleaning or clearing of drains mains services gullies manholes and the like within the site of the Contract Works
consequent upon Damage for which indemnity is provided by this Section

Provided that the Company shall not be liable in respect of seepage pollution or contamination of any Property not insured by this Section.

The reason for the existence of 'debris removal' cover is to allow the insured to recover costs that would otherwise have been outside the protection of the CAR policy. If debris removal is not covered difficulties can arise as to what is an act of debris removal and what is an act of repair. Incidentally clause 22A.4.3 of the JCT contract and similar clauses in 22B and C require the contractor 'to remove and dispose of any debris' and the insurance should cover this.

The expense of removing debris can be very high. There is not only the possibility of rubble from a collapsed building but debris spread over the site following a storm. This rubble and debris cannot be dumped anywhere but may have to be transported to a suitable and permitted place. Note that the insurers' permission for such expense to be incurred must be obtained and it is important to include a sum within item 1 of the schedule (contract works) to cater for such costs.

This cover includes both dismantling and demolishing, shoring and propping up as well as cleaning or clearing of drains and the like, the cost of which can be considerable in the case of flooding of the site. In any event shoring or propping done in order to minimise damage to the works, etc. would be covered and would not need the permission of insurers. The proviso is understandable and merely emphasises that this policy is a material damage one and does not cover third party property or even the insured's extraneous property not within the definition of and listed as insured property in the schedule.

Additional cover 3 – Off-site storage

The indemnity provided in respect of Item 1 of the Property Insured extends to apply to materials or goods whilst not on the site of the Contract Works but intended for incorporation therein where the Insured is responsible under contract conditions provided that the value of such materials and goods has been included in an interim certificate and they are separately stored and identified as being designated for incorporation in the Contract works.

The important point about this cover is that apart from the value to the contractor no CAR policy should be accepted by an architect or engineer on behalf of a client (the employer) without this cover being given in some shape or form, as the JCT 80 contract (as amended) requires it under clauses 16 and 30.3 (see below) and the ICE conditions under clause 54. The clauses of the JCT contract provide as follows:

Clause 16 Materials and goods unfixed or off-site

16.1 Unfixed materials and goods delivered to, placed on or adjacent to the Works and intended therefor shall not be removed except for use upon the Works unless the Architect has consented in writing to such removal which consent shall not be unreasonably withheld. Where the value of any such materials or goods has in accordance with clause 30.2 been included in any Interim Certificate under which the amount properly due to the Contractor has been paid by the Employer, such materials and goods shall become the property of the Employer, but, subject to clause 22B or 22C (if applicable), the Contractor shall remain responsible for loss or damage to the same.

16.2 Where the value of any materials or goods intended for the Works and stored off-site has in accordance with clause 30.3 been included in any Interim Certificate under which the amount properly due to the Contractor has been paid by the Employer, such materials and goods stored shall become the property of the Employer and thereafter the Contractor shall not, except for use upon the Works, remove or cause or permit the same to be moved or removed from the premises where they are, but the Contractor shall nevertheless be responsible for any loss thereof or damage thereto and for the cost of storage, handling and insurance of the same until such time as they are delivered to and place on or adjacent to the Works whereupon the provisions of clause 16.1 (except the words 'Where the value' to the words 'the property of the Employer, but') shall apply thereto.

Clause 30 Certificates and Payments Off-site materials or goods

30.3 The amount stated as due in an Interim Certificate may in the discretion of the Architect include the value of any materials or goods before delivery thereof to or adjacent to the Works (in clause 30.3 referred to as 'the materials') provided that:

30.3.9 the Contractor provides the Architect with reasonable proof that the materials are insured against loss or damage for their full value under a policy of insurance protecting the interests of the Employer and the Contractor in respect of the specified Perils, during the period commencing with the transfer of property in the materials to the Contractor until they are delivered to, or adjacent to, the Works.

Additional cover 4 – Final contract price

This term is probably better known as an escalation clause. It reads:

In the event of an increase occurring to the original price the Limit of Indemnity in respect of Item 1 of the Property Insured shall be increased proportionately by an amount not exceeding 20%

In this way the insurers have built into the policy a 20% inflation factor but this does not protect the contractor and employer from the need to ensure that the limit of indemnity (or sum insured) is always adequate, i.e. of a sufficient sum to cover the cost of repairing any damage which may occur plus debris removal costs and professional fees. Incidentally it is usual to use the term 'limit of indemnity' to apply to the policy limit in liability policies and the term 'sum insured' is used in the case of material damage policies to indicate the policy limit.

Additional cover 5 – Tools, plant, equipment and temporary buildings

The wording of this clause reads:

> The Limit of Indemnity in respect of Items 2, 3 and 5 of the Property Insured is subject to average and if at the time of any Damage the total value of such Item of the Property Insured is of greater value than the Limit of Indemnity the Insured shall be considered as being his own insurer for the difference and shall bear a rateable share of the loss accordingly.

Incidentally, this clause would read more correctly if the letters, 'i.e.' were substituted for the word 'and' in the second line. 'Average' only applies to items 2, 3 and 5 of the property insured as listed in the policy schedule. Once average is applied all claims are affected if the limit of indemnity (sum insured) is inadequate as the claim is scaled down in proportion to the amount the sum insured bears to the full value at risk. Thus, if property worth £40 000 is only insured for £30 000 the insured is under-insured and the amount payable under the policy is calculated as follows:

$$\frac{\text{Sum insured £30 000}}{\text{Value at risk £40 000}} \times \text{Loss sustained}$$

Additional cover 6 – Speculative housebuilding

> The insurance in respect of Item 1 of the Property Insured shall notwithstanding Exception 4(b) for private dwelling houses flats and maisonettes constructed by the Insured for the purpose of sale continue for a period up to 180 days beyond the date of practical completion pending completion of sale.
>
> Practical completion shall for the purposes of this extension mean when the erection and finishing of the private dwelling house are complete apart from any choice of decoration fixtures and fittings which are left to be at the option of the purchaser.

The reason for this additional cover is to protect the insured contractor between the completion of the construction work (in accordance with the second paragraph above) and the sale of the premises to a purchaser for a period of about six months beyond the date of such completion. This is a useful addition which is not usually included in the basic cover of a CAR policy. It is especially useful in relation to theft and malicious damage.

Additional cover 7 – Local authorities

This clause reads:

> The Indemnity provided in respect of Item 1 of the Property Insured shall include any additional cost of reinstatement consequent upon Damage to the Property Insured which is incurred solely because of the need to comply with building or other regulations made under statutory authority with bye-laws of any Municipal or Local Authority

Provided that
1. the Company shall not indemnify the Insured against the cost of complying with such regulations or bye-laws
   (a) in respect of Damage which is not insured by this Section
   (b) if notice has been served on the Insured by the appropriate authority prior to the occurrence of such Damage
   (c) in respect of any part of the Insured Property which is undamaged other than the foundations of that part which is the subject of Damage
2. the Company shall not indemnify the Insured against any rate tax duty development or other charge or assessment arising out of capital appreciation which may be payable in respect of the Property by its owner by reason of compliance with such regulations or bye-laws
3. reinstatement is commenced and carried out with reasonable despatch
4. nothing in this extension shall increase the liability of the Company to pay any amount exceeding the Limit of Indemnity in any one Period of Insurance.

Local authorities have their own bye-laws which allow them to require the owners of property to reinstate damage or destroyed buildings in a way which is acceptable to them. This clause (often included in fire policies as an extension) covers the extra cost incurred by the insured in complying with these requirements. The wording emphasises that the insured cannot claim for the extra cost of improvement he decides to carry out whilst the reinstatement of the damage is being done nor even for changes which are recommended by the authorities. Only when the insured has no alternative but to have the addition or different work done because of the requirements of the local authority, which has the legal power to enforce such requirements, does the clause impose liability on the insurers to pay for this extra expense incurred by the insured.

In the event of this clause being involved in a claim the insurer will obtain confirmation from the authority of the date on which the insured was informed that the extra cost was to be incurred. If this date was prior to the occurrence of the damage resulting in the claim the extra expense is not recoverable from the insurers, even though the notice did not require any alteration or addition to be carried out before the date of the damage. This extra cost and cover is still subject to the limit of indemnity.

The remaining provisos are self-explanatory.

Additional cover 8 – Immobilised plant

The indemnity provided in respect of Items 2 and 4 of the Property Insured shall include the cost of recovery or withdrawal of unintentionally immobilised constructional plant or equipment provided that such recovery is not necessitated solely by reason of electrical or mechanical breakdown or derangement.

This clause draws attention to the fact that although plant may be covered under items 2 and 4 in the schedule there is an exception 3 (see later under the heading 'Exceptions') which excludes damage to plant, tools or equipment due to its own explosion, breakdown or derangement. However, whereas the exception 3 wording only applies to the part responsible and does not extend to other parts of the plant, tools or equipment which sustain direct accidental damage therefrom, this clause excludes the whole of the plant, tools and equipment (including hired-in plant, etc.) immobilised solely by reason of electrical or mechanical breakdown or derangement.

Additional cover 9 – Free materials

Property for which the Insured is responsible shall include all free materials supplied by or on behalf of the Employer (named in the contract or agreement entered into by the Insured)

Provided that the total value of all such materials shall be included in the Limit of Indemnity for Item 1 of the Property Insured and also included in the declaration made to the Company under Condition 2.

As under the majority of construction contracts involving a standard form the employer would be included as an insured in the policy schedule; this wording is again merely for emphasis. However, the proviso should prevent the value of such materials being overlooked when deciding the limit of indemnity for item 1 of the property insured in the policy schedule. The insured must insure the property for its full value (under the JCT form for its reinstatement value).

Condition 2 is a premium adjustment condition when the premium is based on estimates. Consequently it is important to include the value of free materials in making the return at the end of the policy period as required by this condition.

Exceptions

The policy provides that the company shall not indemnify the insured against the following exceptions.

Exception 1 – Faulty workmanship and faulty design

the cost and expenses of replacing or making good any of the Property Insured which is in a defective condition due to faulty defective or incorrect
(a) workmanship

(b) design or specification
(c) materials goods or other property installed erected or intended for incorporation in the Contract Works
but this exclusion shall not apply to accident Damage which occurs as a direct
consequence to the remainder of the Property Insured which is free of such
defective condition.

This exception concerns in (a) and (c) defective workmanship and materials
which insurers consider to be a trade risk. Consequently they will not pay for
the expense of remedying, repairing, or making good defective materials or
workmanship. This policy does not cover the insured's competency in this
respect. However, the exclusion is qualified making it clear that consequential damage to the remainder of the property insured is covered as the
latter damage is free of such defective condition. Nevertheless difficulties
can arise as it is not always possible to indicate what is and what is not a
defective part, and what cost should be excluded.

The meaning of part (b) is clear, namely to exclude the professional negligence risk as the professionals concerned should carry this risk and they
should insure accordingly. If this part of the exclusion did not carry the
later qualification (as is the case with some policies) it would dispose of
most difficulties but with the qualification which excludes only the defective part there can be controversy in identifying the part which is defectively designed. For example defective piling can result in a complete
rebuild.

The JCT exclusion
The JCT contract, by clause 22.2 in the 1986 amendment, set outs the
meaning of 'all risks insurance' in that contract and the same type of
exclusion reads as follows:

> any work executed or any Site Materials lost or damaged as a result of its own
> defect in design, plan, specification, material or workmanship or any other work
> executed which is lost or damaged in consequence thereof where such work relied
> for its support or stability on such work which was defective.

Now this exclusion is wider than the exception under discussion as it
excludes 'work [which] relied for its support or stability on such work which
was defective' as well as excluding the defective part. So if a beam supporting a ceiling is held up by a defective bolt and the beam collapses
because of the defective bolt and the beam falls through the two floors
underneath, there is damage to the bolt, to the beam and to the floors (all part
of the contract works). The exception under discussion, as it only excludes
the defective part, would only exclude the bolt, but the JCT exclusion would
exclude the bolt and the beam as the latter relied for its support on the
defective bolt.

This indicates that the policy exception under discussion would only
exclude work which is not of much value to remedy, i.e. the bolt, whereas the
JCT form would exclude both the cost of the beam replacement as well as the

bolt which is more expensive. Neither exception would exclude the damage to the floors below, which would be the major expense involved.

The Pentagon Construction *case*

The Canadian case of *Pentagon Construction (1969) Co Ltd* v. *United States Fidelity and Guarantee Co* (1978) illustrates the difficulties in differentiating between defective workmanship and defective design, although that case did not concern the limited cover given by UK insurers for defective material or workmanship. The interpretation of the following wording was considered.

'This insurance does not cover:
(a) Loss or damage caused by:
    (i)   faulty or improper material or
    (ii) faulty or improper workmanship or
    (iii) faulty or improper design.'

In this case Pentagon was a building contractor engaged to construct a sewage treatment plant which included a concrete tank. The plans and specifications required a number of steel struts to be laid across the top of the tank with each end welded to a plate let into the concrete wall beneath it. The purposes of the struts were to strengthen the tank by holding the sides together and to hang equipment from them. The contract required Pentagon to test the tank. Pentagon insured the work under a contractors' all risks policy.

After the concrete work of the tank was completed and the struts laid across the tank, but before the end of the struts had been welded, the tank was tested by filling with water. The tank bulged and a claim was lodged under the policy and repudiated by the insurers who relied on the above exclusion.

At the court of first instance it was held that the design of the tank was not faulty or improper and there was no faulty or improper workmanship. The insurers appealed, and argued that the word 'design' included the plans and specifications and that they were faulty in that they omitted to state that the tank should not be tested until after the struts had been welded.

All three judges in the appeal court decided that the evidence clearly established that the wall of the tank failed because of the failure to weld the steel struts to the top of each side of the wall before testing; this amounted to improper workmanship. This led one judge to decide it was unnecessary for him to consider the question of faulty or improper design. The other two judges reached different opinions on the meaning of 'design'.

The conclusions to be drawn from this case are that:

● Workmanship is not limited to the work or result produced by a worker. It includes the combination or conglomeration of all the skills necessary to complete the contract, including, in this case, the particular sequence necessary to achieve the performance of the contract. Failure to follow that sequence could constitute faulty or improper workmanship and in this case did so.

- It is not known whether:
  - (i) detailed instructions on the plans and specifications on how the work of construction is to be carried out are not part of the design, which was one judge's view (he added that if he were wrong he did not think it was necessary for the plans and specifications to warn that the tank should not be filled with water before the struts were welded); or
  - (ii) design includes the drawings and specifications, which was the other judge's view.

Thus on the meaning of 'design', with the third judge abstaining, the case is unsatisfactory. Defective workmanship is a contract hazard normally accepted by contractors. The cost of doing such work twice is not properly a matter for insurers (but the cost of rebuilding other parts of the insured property is), and if the use of defective materials is not due to the negligence of the contractor, he will probably have a remedy against the suppliers.

The Queensland Government Railways *case*
Another case on the meaning of 'faulty design' as used in the exclusion to this policy is *Queensland Government Railways and Electric Power Transmission Pty Ltd* v. *Manufacturers Mutual Insurance Ltd* (1969). A railway bridge in Australia was being constructed by Electric Power Transmission Pty Ltd for Queensland Government Railways, railway authority, to replace the bridge built in 1897 which had been swept away by flood waters. Prismatic piers (similar to the original piers, but strengthened) were being erected when they were overturned by flood waters after exceptionally heavy rains. EPT and QGR claimed to be indemnified by the insurers under a contractors' all risks policy, which provided, *inter alia*:

'... this insurance shall not apply to or include:
(vii) cost of making good faulty workmanship or construction...
(xi) loss or damage arising from faulty design and liabilities arising therefrom.'

The insurers denied liability, contending that the loss was due to faulty design of the piers. The arbitrator found that, in the state of engineering knowledge at that time, the design of the new piers was satisfactory. Investigations into the cause of failure of the piers showed that during floods they were subjected to greater transverse forces than had been realised, nevertheless the loss was not due to faulty design, in that 'faulty design' meant that 'in the designing of the piers there was some element of personal failure or non-compliance with the standards which would be expected of designing engineers'. Therefore, the insurers were liable. They applied to have the award set aside or remitted on the ground that the arbitrator misconstrued the term.

It was held by the Supreme Court of Queensland that, in the context, 'faulty design' implied some element of blameworthiness or negligence, which had been negatived by the arbitrator's findings; that subsequently acquired knowledge revealing that the piers were not strong enough could not convert the design, which would at the time have been accepted by responsible and competent engineers, into a 'faulty design', and that, therefore, the insurer was liable.

On appeal the decision of the Queensland Supreme Court was reversed. It was held that 'faulty design' did not imply an element of blameworthiness or negligence; the loss of piers through the inadequacy of their design to withstand an unprecedented flood was outside the policy, notwithstanding that the design complied with the standards that would be expected of designing engineers according to engineering knowledge and practice at the time of their design.

The Kier Construction *case*

In *Kier Construction* v. *Royal Insurance (UK)* (1994) the plaintiff had agreed with the Central Electricity Generating Board (CEGB) to build water-cooling works for a power station. CEGB had agreed to take out a single site insurance policy on the works on behalf of itself and all contractors and sub-contractors, which they did. The type of soil damaged the piles and as a result the supported structure was not watertight as required. The plaintiff claimed under the policy, which the insurers rejected alleging, *inter alia*, that the damage to the piles came within the phrase 'defective in workmanship', which was an exception under the policy. (Other allegations will be dealt with elsewhere in this book.)

It was held that the phrase 'defective in workmanship' had to be construed objectively and did not involve only an enquiry into whether or not there was negligence (see the *Queensland Government Railways* case mentioned earlier). Furthermore, on a true construction of the contract documents, ground disturbance was not intended to be a matter of defective workmanship and therefore not intended to fall within the exclusion. In any event, on the facts defective workmanship had not been shown. This is significant as some commentators had considered that as a result of the *Queensland Government Railways* case the mere fact that a structure failed to meet all weathers, however extreme, rendered it of faulty design without more.

The Hitchins *case*

In *Hitchins (Hatfield) Ltd* v. *Prudential Assurance Co Ltd* (1991) the plaintiff, a firm of builders, was insured under a contractor's policy issued by the defendants. The policy covered loss arising from 'any fault, defect, error or omission in design', but went on to exclude liability for 'any increased costs due to redesigning the property insured or any part thereof which is defectively designed'.

The plaintiff was engaged in building a housing estate on a slope, which he levelled off to create four separate terraces but increased the slope between the terraces. The slope was unstable due to the composition of the soil and landslips occurred. The Prudential relied on the exclusion to avoid payment to cover the cost of reinstating the slope by arguing that the slope had been 'defectively designed'. This point came before the Court of Appeal as a preliminary issue. It was agreed by both parties that the words 'defect in design' covered both negligent and non-negligent defects in design in accordance with the *Queensland Government Railways* case mentioned earlier. The plaintiff argued that the meaning of the words 'defectively designed' in the exclusion clause imported an element of negligence, which meant that the Prudential had to demonstrate the plaintiff's negligence to rely on the exclusion. The Prudential argued that there was no difference between the phrases 'defect in design' and 'defectively designed' so the loss was excluded from liability.

The Court decided that the two phrases had to be distinguished and that a building could have a defect in design with nobody at fault, but 'defectively designed' referred to the conduct of the designer. The distinction was between an inanimate object (defect in design) and the activity which gave rise to that inanimate object (defectively designed). Furthermore it was clear that some distinction had to be made, as the policy deliberately used two different phrases and had to be presumed to be referring to two different concepts. Thus the Prudential could rely upon this exclusion only by showing that the plaintiff had been negligent. However, it should be noted that the Court of Appeal has confirmed the decision that a 'defect in design' can occur without any negligence upon the part of the designer.

The BC Rail *case*
A similar problem arose in *BC Rail Ltd* v. *American Home Assurance Co* (1992) heard by the British Columbia Court of Appeal. Here work between 1983 and 1985 to support a railway line on a steep embankment, designed by an employee of the plaintiff, failed due to inadequate soil tests, and a landslide occurred. BC had to reroute rail traffic (costing $456,340), pending the building of a temporary bridge (costing $598,210), and eventually a permanent bridge (costing $1,263,707).

The wording of the BC all risks policy defined the perils insured against as 'all risks of physical loss or damage from any cause except as herein after excluded'. The policy excluded loss or damage caused by 'error in design – however, damage resulting from ... error in design ... is hereby covered'. The questions were had there been an error in design and if so was it possible to reconcile this apparently contradictory wording? The view of the majority of the court was that the word 'error' in 'error in design' referred to the design itself and not to the workmanship producing the design. Thus whether the BC employee was negligent or not was irrelevant and the exclusion applied to the loss. The dissenting judge considered the word 'fault' in the previous cases discussed earlier was not the same as 'error'. 'Fault', he thought, could refer either to the design or to the workmanship itself, but 'error' could refer only to the workmanship. Thus, he ruled that the insurer could not exclude the claims without proof of negligence on the employee's part.

Turning to the second question mentioned above the majority of the court held that the costs of reconstruction (the two bridges) were not 'damage resulting from' the error within the meaning of the exclusion clause, as the railroad bed was an integral part of the design. So this clause operated to exclude these costs. The expenses of rerouting traffic were not 'physical loss or damage' within the meaning of the primary definition (the operative clause), and the references to 'loss or damage' and 'damage' in the exclusion clause should be construed to mean physical damage only. The rerouting costs were economic loss and not covered by the policy. Consequently, all BC Rail's claims failed.

Apart from the decisions on design this case also indicates the importance to insurers of not merely excluding consequential loss from all risks policies, but to use the word 'physical' in the operative clause in order to restrict the cover in this respect.

Drafting points
If there are any conclusions to be drawn from these cases the following are suggested to those drafting exclusions in policies:

(1) Forget style. You are not writing a best seller or going for a Nobel prize. Once senior courts have decided on the meaning of a phrase and you intend to convey that meaning, keep on using it. However repetitive it appears, stick to it. Good English in this respect is a secondary consideration. Any change in phraseology may result in a court deciding you meant to change the meaning, even though you were only thinking of making it more readable and had no intention of changing the meaning. It could be an error that could cost your insurer employer dearly (see the *Hitchins* case).

(2) Do not consider the dominant phrase only. Even supporting phrases can make all the difference in the dominant phrase. For example, the effect of the decision in the *BC Rail* case was that the words 'damage caused by' and 'damage resulting from' when applied to 'error in design' meant that the former phrase concerned the design itself and the latter did not. The reasoning was that the defects (the design of the railroad bed) were within the policy exclusion, as otherwise the policy would amount to a warranty as to the quality of the product and insurance policies only covered unforeseen risks. The policy thus excluded all costs incurred in correcting the defect which disposed of the claim for the building of the temporary and permanent replacement bridges.

Exposure work and other matters
Nothing is said in the policy about the cost of dismantling or exposure work necessary to get at a defective part. It is arguable both ways, i.e. it is part of the cost of rectifying the property excluded by the policy or it is part of the insured property not excluded. The same position can arise when a limited

merely emphasises the basic position under the policy which is stated elsewhere.

(e) This exclusion is of existing structures which are not part of the contract works and clearly not covered by the policy. Nevertheless what is not always clear is what are works and what are existing structures when there is an extension of an existing structure or an alteration of an existing structure.

In the JCT 80 edition the 'Works' are defined in clause 1.3 as those 'briefly described in the First recital and shown and described in the Contract Drawings and in the Contract Bills'. Loosely, 'Works' in the same contract means the work done by the contractor and not yet handed over and also the unfixed materials and goods, delivered to, placed on or adjacent to the work done and intended for incorporation therein in accordance with clause 22. On the other hand in the JCT contract clause 22C.1 the existing structures which are the responsibility of the employer as well as the contents thereof owned by him must be insured. Therefore although the employer is required by that clause 22C.1 to take out a joint names policy covering those responsibilities against specified perils it is clear that the wording under discussion would not cover the existing structures.

(f) Both parts of this exception seem almost unnecessary as it can only emphasise what is already stated in the policy schedule.

Exception 4 – Damage after completion

Damage to the Contract Works or any part thereof
(a) caused by or arising from use or occupancy other than for performance of the contract or for completion of the Contract Works by or on behalf of the Insured
(b) occurring after practical completion or in respect of which a Certificate of Completion has been issued unless such Damage arises
   (i)   during any period (other than the Maintenance Period) not exceeding 14 days following practical completion or issue of such Certificate in which the Insured shall remain responsible under the terms of the contract for the Contract Works or the completed part thereof
   (ii) during the Maintenance Period and from an event occurring prior to the commencement thereof
   (iii) by the Insured in the course of any operations carried out in pursuance of any obligation under the contract during the Maintenance Period.

This exception concerns loss or damage after the contract works (or any part thereof) have been completed and delivered to or taken into use or occupation by the principal (employer), except to the extent that the insured may remain liable:

● under the maintenance conditions of the contract;
● during a period not exceeding fourteen days after the issue of a certificate of completion and which is the responsibility of the insured to insure.

So far as the JCT 80 form is concerned, sub-clause 22A requires the contractor to insure until 'the date of issue of the certificate of Practical Completion'. In any event, once the works have been completed and delivered, it is arguable that the contractor no longer has an insurable interest in the works, assuming he has been paid and subject to the contract terms. Sometimes insurers are asked to cover occupation before completion, and they may be prepared to do so for an additional premium.

Any occupation will usually involve a change of information on which the insurance business was placed and thus a probable change of rate for premium purposes. In case of failure to notify the insurer, the fact of the employer's or his tenant's occupation is rarely in doubt but the part of the property occupied can be in contention. However, once notification takes place, in most cases the problem is resolved quite easily. At practical completion the risk of loss or damage passes to the employer who is responsible for arranging his own insurances. Clause 18 of the JCT 80 form deals with 'partial possession by the employer' and reads as follows:

18   Partial possession by Employer

18.1 If at any time or times before the date of issue by the Architect of the certificate of Practical Completion the Employer wishes to take possession of any part or parts of the Works and the consent of the Contractor (which consent shall not be unreasonably withheld) has been obtained, then notwithstanding anything expressed or implied elsewhere in this Contract, the Employer may take possession thereof. The Architect shall thereupon issue to the Contractor on behalf of the Employer a written statement identifying the part or parts of the Works taken into possession and giving the date when the Employer took possession (in clauses 18, 20.3, 22.3.1 and 22C.1 referred to as 'the relevant part' and 'the relevant date' respectively).

18.1.1 For the purposes of clauses 17.2, 17.3, 17.5 and 30.4.1.2 Practical Completion of the relevant part shall be deemed to have occurred and the Defects Liability Period in respect of the relevant part shall be deemed to have commenced on the relevant date.

18.1.2 When in the opinion of the Architect any defects, shrinkages or other faults in the relevant part which he may have required to be made good under clause 17.2 or clause 17.3 shall have been made good he shall issue a certificate to that effect.

18.1.3 As from the relevant date the obligation of the Contractor under clause 22A or of the Employer under clause 22B.1 or clause 22C.2 whichever is applicable to insure shall terminate in respect of the relevant part but not further or otherwise; and where clause 22C applies the obligation of the Employer to insure under clause 22C.1 shall from the relevant date include the relevant part.

18.1.4 In lieu of any sum to be paid or allowed by the Contractor under clause 24 in respect of any period during which the Works may remain incomplete occurring after the relevant date there shall be paid or allowed such sum as bears the same ratio to the sum which would be paid or allowed apart from the provisions of clause 18 as the Contract Sum less the amount contained therein in respect of the relevant part bears to the Contract Sum.

It is arguable that 'possession' in this clause is not the same or may not be the same as 'caused by or arising from use or occupancy' which is the expression used in the policy, as merely use for storage may not involve taking possession. Nevertheless, such use would fall foul of the policy wording.

Secondly, the question arises as to the effect of such use or occupancy. Does it just affect the cover for that part of the site used or occupied, or is the whole of the policy cover prejudiced? Probably the answer depends on the additional risk the use or occupance imposed on the remainder of the site which is not used or occupied. If the whole site risk is increased then it will affect the whole policy cover. For example the storage of inflammable material could affect the whole site.

The responsibility to insure in the case of sectional completion was considered in *English Industrial Estates Corporation* v. *George Wimpey & Co Ltd* (1973) although this was a case of *alleged* sectional completion.

A factory owned by English Industrial Estates Corporation was leased to Reed Corrugated Cases Ltd. In 1969 Reed wanted to extend the factory, while continuing to make corrugated cardboard, to install a large new machine and have storage space for hundreds of reels of paper.

Wimpey tendered for the contract and was successful. The tender was on the pre-1980 JCT form incorporating the bills of quantities. Wimpey had a blanket contractors' all risks policy which provided the cover required by sub-clause 20(A)(1), now sub-clause 22A.1 under JCT 80.

By January 1970, a great deal of the work had been carried out by Wimpey, and Reed had installed the machine and had stored 1500 reels in the new reel warehouse. On 19 January 1970 much of the new factory was gutted by fire and the damage was estimated at £250 000. At that time the contractors had not finished their work.

The Corporation argued that, as the works had not been completed, it was Wimpey's duty to insure, and that the loss, therefore, should fall upon it or its insurers.

Wimpey, however, argued that the Corporation, through Reed, had taken possession of several parts of the works and that, under clause 16, the risk had passed to the Corporation in respect of those parts. For the wording of clause 16 see clause 18 of JCT 80.

The question before the court was, had the employer before the date of the fire 'taken possession' of any part or parts of the works? If it had, that part was at the employer's risk. If it had not, then the contractor was at risk.

The facts showed that the car park, for example, was accepted for handover by the employer's architect in September 1969, when he issued a certificate on a RIBA printed form which certified that 'a part ... of the works, namely, car park, the value of which I estimate to be £10,000 was completed to my satisfaction and taken into possession on 22 September 1969...'

An architect gave evidence that that form of certificate was normal for sectional completion.

The matter was complicated by several provisions in the bills of quantities showing that Reed would install plant and equipment and would occupy and use part of the works, but that the contractor was still to keep the works covered by insurance. Lord Denning was prepared to consider clause 16 without placing any reliance on the provisions just mentioned. In his opinion the words 'taking possession' of a part of the works must be so interpreted as to give precision to the time of taking possession and in defining the part, because of the important consequences which followed on it. In any event clause 2.2 states that nothing contained in the contract bills shall override or modify the application or interpretation of that which is contained (*inter alia*) in the conditions.

To achieve this precision, the parties themselves had evolved suitable machinery to determine it by way of a definite handing over of the part by the contractors to the employers. The practice was for the contractors to tell the architect that a party was ready for hand-over. The architect would inspect it and, if satisfied, would accept it on behalf of the employers. He would give a certificate defining the part, its value and the date of taking possession. The hand-over was thus precise and definite. It was the accepted means of defining the hand-over.

In Lord Denning's view, the contractor at the time of the fire had not handed over to the employer the responsibility for the board machine house, the reel warehouse or the two extensions. Although Reed was using those places, it was the contractor's responsibility to insure them until actual hand-over. The risk remained with it and its insurers must bear the loss.

The other two judges in the Court of Appeal came to the same conclusion as Lord Denning, but they merely considered that some formality was required in interpreting clause 16.

Clearly there are dangers for contractors in allowing use or possession of the works which they are insuring for all risks under clause 22A without clarifying responsibility for insurance of their part of the works concerned. However, if, before practical completion of any part of the works is reached, and consequently clause 18 could not apply, the employer requires to use part of the works, then this should not be allowed without the insurance position being agreed. Incidentally, it is just as important that the party responsible for the insurance of the partly occupied premises should agree the position with their insurers.

It seems from the above case that whatever type of construction contract applies some clear cut and final formalised conduct by the parties must be shown, evidencing the transfer of possession so that the parties are in no doubt that the various consequences of taking possession have come into effect. Care must be taken to deal with the various consequential matters including, and particularly, those relating to insurance.

Exception 5 – Risks excluded by the contract

> Damage for which the Insured is relieved of responsibility under the terms of any contract or agreement.

In some construction contracts the contractor is relieved of responsibility for loss or damage to the works in certain circumstances. Thus, under the optional clauses 22B and 22C of JCT 80 (as amended in 1986) the employer is responsible for the works and site materials (as defined), in that he is to take out and maintain a joint names insurance policy (in the joint names of the main contractor and the employer) for all risks insurance as defined in clause 22.2.

Under the sixth edition of the Institution of Civil Engineers Conditions of Contract the 'excepted risks' (those for which the main contractor is not responsible) so far as the contract works is concerned, include damage due to use or occupation by the employer, his agents, servants or other contractors (not employed by the main contractor), or due to fault, defect, error or omission in the design of the works (other than a design provided by the contractor pursuant to his obligations under the contract) as well as war and kindred risks, nuclear risks, and sonic waves.

Exception 6 – Liquidated damages and consequential loss

> (a) liquidated damages or penalties for delay or non-completion
> (b) consequential loss of any nature

Regarding (a) insurers prefer to work on an indemnity basis when paying claims under a material damage policy and are reluctant to enter into payments on an agreed value basis. Apart from this, penalty payments as defined are consequential losses and the following remarks apply.

Turning to (b) the operative clause of the CAR policy is sometimes worded so that without this exclusion it would cover consequential loss. The danger for the insurer is that legally 'consequential loss or damage' has been held to mean loss which does not result directly and naturally from the act concerned or, in the situation under discussion, the perils covered by the policy. The point is that the type of loss which (within the insurance industry) is considered by insurers (in their terminology) to be consequential loss is not in fact considered by the courts to be consequential because it is a direct and natural result, e.g. loss of profit from loss of use, increased cost of working, and loss arising from delay in completing contracts. Therefore, a safer way is for the insurer to state specifically what he or she intends to exclude by the words 'consequential loss', otherwise the legal interpretation of 'indirect or consequential' will not include those heads of damage intended by the insurer to be excluded from the policy cover of the CAR policy. However, this could result in a lengthy list, which even then may not exclude everything intended to be excluded from the policy. Possibly the best way is to confine the operative clause to physical or material loss or damage, which the policy in Appendix 2 does. For the legal authorities on this aspect see the line of cases from *Millar's Machinery Co Ltd* v. *David Way & Son* (1934) to *Croudace Construction Ltd* v. *Cawoods Concrete Products Ltd* (1978). A combination of both methods may be even more satisfactory to all concerned, specifically excluding the financial losses mentioned above.

Some consequential losses may be covered under special policies; sometimes the employer (principal) is the insured. These policies normally give cover against the same perils as the material damage policy required for the protection of the works and are subject to such a policy operating. Consequential loss policies can be obtained covering the following costs.

Advance profits
The protection provided by this policy is in respect of financial loss through delay by damage to the works or at the supplier's premises of important plant or equipment or during transit. Payment under this policy does not start until the date the business would have commenced but for the damage, and is in respect of the anticipated income, i.e. the gross profit of a manufacturer or the rent of a property developer which is not earned at the estimated date.

Additional cost of working
This is an expense incurred by the contractor and can be an extension of the contractors' all risks policy as it follows a claim under that policy and involves costs beyond those incurred in making good that damage. For example, the basis of claims settlements in the case of payment to workmen by reason of guaranteed time or such agreements is by calculating the difference between the amount paid to the workmen and the amount which would have been payable had no such agreement been in force. In the case of plant standing idle, the calculation for hired plant as the amount payable for the affected period and for the contractors' own plant, is based on an allowance in respect of loss of working time, say 66% of the rates for such plant in either of the publications applicable (*Definition of Prime Cost of Daywork carried out under a Building Contract*, published jointly by the RICS and BEC (formerly the NFBTE), or the *Schedules of Dayworks carried out incidental to Contract Work*, issued by the Federation of Civil Engineering Contractors). Sometimes the expense is difficult to identify, especially when time has also been lost on the contract before the damage.

Fines and damages
This is also a possible extension to the contractors' all risks policy covering fines and damages payable by the contractor under the construction contract, following loss or damage due to some or all of the perils covered by this all risks policy. This cover is not easily obtainable and the rate of premium is high.

Additional or extended interest charges
Where the subject matter of the contract is to be sold on completion, in the case of damage there could be delay in receipt of the money from the sale. The cover can give the agreed amount of interest on the new amount of the sale to the extent that it is delayed subject to a limitation of the indemnity period. On the other hand, the actual interest on a loan could be insured so far as it is extended due to the damage.

Exception 7 – Sonic waves

> Damage occasioned by pressure waves caused by aircraft or other aerial devices travelling at sonic or supersonic speeds.

Material damage and loss of profits policies were never intended to cover loss, destruction or damage directly occasioned by sonic waves. However, sub-clause 1.3 of JCT 80, as amended, defines the Excepted Risks as including sonic waves.

Furthermore, the UK Government indicated that if such damage were to result from Concorde's flights it would pay compensation. In fact this does not appear to have been necessary. The material damage policies concerned would include not only the CAR policy but also the specified perils policy mentioned in Chapter 12.

Exception 8

> the Excess specified in the Schedule.

The CAR policy can have an overall excess of £250 or more, however the loss or damage arises, and/or a higher excess for whatever the underwriter considers the more hazardous perils.

Exception 9

> Damage in Northern Ireland caused by or happening through or in consequence of
> (a) civil commotion
> (b) any unlawful wanton or malicious act committed maliciously by a person or persons acting on behalf of or in connection with any unlawful association
> For the purpose of this exclusion
> > (i) unlawful association means any organisation which is engaged in terrorism and includes any organisation which at the relevant time is a proscribed organisation within the meaning of the Northern Ireland (Emergency Provisions) Act 1973
> > (ii) terrorism means the use of violence for political ends and includes any use of violence for the purpose of putting the public in fear
> In any suit action or other proceedings where the Company alleges that by reason of this Exception any Damage is not covered by this Section the burden of proving that such Damage is covered shall be on the Insured.

This is an insurance market clause the reason for which is obviously as a result of the unrest in Northern Ireland. It should be noted that whereas normally the burden of proving that an exception applies is on the insurer, in this particular exception the burden (of proving that the exception does not apply) is placed on the insured.

Terrorism cover

Terrorism losses became so serious on the mainland of Great Britain, especially following the first 'City' explosion, initially reported as a loss of some £800 million, that reinsurers withdrew their reinsurance for direct writing insurers in the UK.

In April 1994 the JCT published a *Guide to Terrorism Cover* which provides amendment provisions in the JCT forms on insurance of the works and of the existing structures and contents for physical loss or damage due to fire or explosion caused by terrorism. However, these arrangements effective from 1 July 1993 only apply to 'Mainland' Great Britain, i.e. they do not include Northern Ireland, the Channel Islands or the Isle of Man.

Appendix 12 is an extract from 'Part A Introduction' to the JCT Publication mentioned above. The Amendments referred to in paragraph 6 of Appendix 12 are not intended to be a permanent amendment of the JCT Forms to which they relate and will thus not be incorporated in reprints. Nevertheless they should be adopted while the following problem remains; namely where through no fault of either Employer or Contractor, but by a Government decision over which neither party has any control, the Works and any existing structure may be deprived of terrorism cover for fire and explosion damage while the Works are in progress. The Amendments therefore provide alternatives for the Employer to choose from if such termination of terrorism cover were to occur.

Using the works insurance provisions in JCT 80 as an example the amendments have to deal with two situations depending on whether the Contractor (clause 22A) or the Employer (clause 22B or clause 22C.2) is required to take out and maintain All Risks cover for the Works.

On an *existing contract* the parties may wish to include in their contract the relevant Amendment referred to in the Guide. A supplementary agreement, executed under hand or as a deed, according to the method of execution of the existing contract, would be necessary to give effect to any such decision by the parties. A model agreement is set out in the Guide.

As this terrorism cover came into force largely due to the activities of the IRA and was shortly followed by a peace initiative, one wonders whether it is a thing of the past. Only time will tell.

Other provisions

Extensions

The only extension applying to the CAR policy is the indemnity to principal extension and this was dealt with in Chapter 6.

General exceptions

The nuclear risks, the contractual liability and the war risks exceptions have already been considered in Chapter 6.

Conditions

The only condition which has not been discussed already in Chapter 6 is Condition 11 (Rights). It reads as follows:

> In the event of Damage for which a claim is or may be made under Section 3
> (a) the Company shall be entitled without incurring any liability under this Policy to
>   (i) enter any site or premises where Damage has occurred and take and keep possession of the Property Insured
>   (ii) deal with any salvage as they deem fit
>   but no property may be abandoned to the Company
> (b) if the Company elects or becomes bound to reinstate or replace any property the Insured shall at their own expense produce and give to the Company all such plans and documents books and information as the Company may reasonably require. The Company shall not be bound to reinstate exactly or completely but only as circumstances permit and in reasonably sufficient manner and shall not in any case be bound to expend in respect of any one of the items of Property insured more than the Limit of Indemnity in respect of such item.

In a number of aspects this seems to be a 'belt and braces' condition in that it merely emphasises rights which would exist without this condition.

Thus in respect of (a), while the onus of proof is primarily on the insured, the insurers could hardly be expected to pay a claim without investigation, which must involve entering the site and at least examining the property insured. However, the abandoning of property to the insurers is not a situation the insurers would want as they might become liable for potential third party liabilities, cost of removal, etc. In any event, the insured's claim is basically for the loss of value, i.e. after salvage value has been considered.

Similarly in respect of (b), the insurers are entitled to all information to support a claim which they may reasonably require, and the insurers in complying with the principle of indemnity would not be expected to reinstate exactly or completely but only as circumstances permit and in a reasonably sufficient manner up to the limit of indemnity stated in the policy schedule.

7.2 *Types of policy cover*

Single contract policy

The contractors' all risks policy can be arranged on a 'contract by contract' basis. However, this may be considered undesirable because of the danger of overlooking the necessity to insure, and also in view of the greater administrative work involved in costing insurance for each contract. Another danger of this type of policy is that cover ceases when the policy terminates. The contractor therefore needs to have a separate 'residual risks' policy (a separate annual policy to cover minor works, including going back to work on lapsed contracts). On the other hand this type of cover gives the underwriter more details of the insured's method of construction.

Blanket (or floater) policy

Contractors frequently find it prudent to arrange this type of contractors' all risks policy, particularly where a considerable number of contracts are undertaken each year. This policy is renewable annually and charged at fixed premium rates. Although there is usually an exclusion of hazardous operations, e.g. foundation work on existing structures, bridges, viaducts, tunnelling, work in water etc., the policy has the following advantages:

- As the cost is known, tendering is simpler.
- There is a minimum number of policies and minimal administrative work.
- There is automatic cover on almost if not all contracts, i.e. without notification to insurers.
- An estimated premium is paid at the beginning of each year and that premium is adjusted at the pre-fixed rate at the year end according to the total insured sums involved.
- The policy can be extended to cover all constructional plant and equipment used on all the sites throughout the year.
- The policy is usually issued to a contractor to insure him, and his employer and sub-contractors also, in respect of accidents occurring in connection with the work described in the policy. A blanket policy may also be issued to an industrial, commercial or other type of employer who frequently employs contractors for construction, extension or alteration of its own property.
- Underwriters can monitor claims and adjust premiums.

7.3 *Claims presentation*

Some contractors tend to perform poorly in their method of claims presentation under the CAR policy. CAR claims should be presented under clear and precise headings of claim supported by authenticated daywork sheets or other approved site documents and plant and material allocation forms which will enable claim settlements to proceed smoothly. Poor presentation can involve delays due to misunderstandings with insurers or loss adjusters.

The remedial works must be the subject of as detailed a specification as possible and a bill of quantities. At the earliest possible stage daywork areas must be indicated and a basis of costing agreed. Valuation as 'daywork' arises when additional or substituted work cannot be valued by measurement. See clause 13.5.4 of JCT 80 edition or clauses 52(3) and 56(4) of the ICE conditions.

The cost of remedial works is dependent, *inter alia*, on the time involved. Thus it is necessary to have a programme of works arranged from the start. This may involve site demolition and clearance before the remedial works proper can commence. Specialist sub-contractors may undertake this type of work as it may be more economic from a direct cost viewpoint, leaving the main contractor free to concentrate on work on other areas of the contract.

Under JCT 80 as amended in 1986 (see Chapter 11) by clause 22A.1 the insurance cover must be for the full reinstatement value of the works plus a percentage for professional fees, and while before 1986 temporary buildings, plant, tools and equipment were specifically excluded it must be assumed that this is still the position. In any event these other items of property, e.g. temporary works, constructional plant, scaffolding, tools, equipment, site huts and other temporary buildings in practice will also be insured by the contractor.

Therefore, under clause 22A.1, bearing in mind that the contractor also still has his obligation under clause 2.1 to carry out and complete the works in the contract documents, the problem of adequacy of insurance becomes the contractor's problem. If either of the optional clauses 22B or 22C apply then the insurance of the works becomes the employer's responsibility but more detail of these requirements will be found in Chapters 11 and 13.

It is important for contractors to keep a record of any construction which is demolished in terms of dimension, quality and photographs which need to be dated. A few important records of this kind taken at the right time can do wonders in substantiating the veracity of a claim.

The costs submitted, so far as they relate to the repair or reinstatement of the permanent or temporary works, will consist basically of labour costs, material and plant costs and overheads. Incidentally it is usually the loss adjuster who recommends the payments on account. The payment being in the joint names of the employer and contractor, subject to the wording of the construction contract concerned. One of the most invidious situations on cost arises from the extent to which preliminaries should be added to labour, material, plant and overhead costs. It very much depends upon how the contractor has priced the bill and indeed, how the bill of quantities has been formulated as to whether the contractor ought reasonably be entitled to a certain proportion of preliminaries.

Labour costs

The labour costs are supported by daywork sheets which record the type of labour employed, the number of hours entered for each class of labour, usually the tasks performed, a rate and the extent to which that rate is varied by on-costs such as bonuses, subsistence or lodging, travelling, etc., either individually or on a group basis. The day work sheets should be authorised by a responsible official on site and/or should as far as possible have been approved by the loss adjuster during interim site visits.

On general building work there can be two methods of calculating the main contractor's labour costs. The first is the unit rate included in the original main contract bill of quantities, used for tender purposes. This method can be used where the damaged areas are easily recognised as part of the original bill. Where the items of damage were not included in the original bill there is no alternative but to adopt a daywork basis or agree a unit figure. In the case of hourly-paid labour, questions sometimes arise concerning waiting time for either the arrival of materials, or plant, or just an improvement in the weather. Provided the labour has been properly and reasonably brought to the site and cannot be employed elsewhere either on-site or at other nearby sites such costs are usually recoverable. There is also a possibility of waiting time in respect of instructions which might need to be given by the designer, supervisor, engineer if an unforeseen problem arises.

Overtime working would not normally be paid by insurers on the grounds that the insured benefits in that he can avoid other liabilities, but in some circumstances such overtime has the effect of reducing on-costs such as travelling and lodging by reducing the repair period overall and this may be recoverable under the policy. Also there may be other acceptable reasons why overtime payments should be allowed.

A method of calculation of hours worked and the calculation of labour costs is published annually by the Building and Allied Trades Joint Industrial Council (BATJIC). There are also complementary wage awards applicable with other allied industries such as mechanical, electrical, heating and ventilation engineering.

Material costs

The cost of materials should be straightforward provided that the nature and amount of the materials is in accordance with the particular loss or damage. The claim must be supported by invoices. The main contractor is permitted to charge the invoiced price of all materials after deducting VAT including delivery to site and accounting for trade discounts.

Major changes to the VAT status of supplies of goods and services by contractors came into effect on 1 April 1989. It is clear that on and from that date the majority of work supplied under contracts on JCT contracts will be chargeable at the standard rate. The exception is private housing. The VAT status of the employer will have a bearing on the declared contract value and possibly the valuation of claims.

The price adjustment formulae for up-grading prices to current value were originally set up 25 years ago under the auspices of the National Economic Development Office and are still familiarly referred to as 'the NEDO Indices'. The first working groups were chaired by Mr J.W. Baxter and Mr J.G. Osborne whose names are still linked to the indices.

The *Monthly Bulletin of Indices* currently published by the Department of the Environment is used in conjunction with formula methods for adjusting resource costs to allow for variation of price. The formula method has been widely used for nearly 25 years both in the public and private sectors. It allows tenderers to price work with confidence in the knowledge that they will receive speedy reimbursement of fluctuations in cost. As from December 1994 this Bulletin has been redesigned to accommodate a new series of indices based at 1990 covering all categories of work. However, the previous Series 2 Indices for building and specialist engineering works and the 1970 Series of Indices for civil engineering works will continue unaltered for the foreseeable future. The *Monthly Bulletin* can be obtained from HMSO quoting ISSN number 0964-4575. These statistics are used to act as multiplication factors to change the rates quoted for particular items, in any specific category of building work from the date of origin to the date on which the work was carried out.

Overheads

The most common overheads are site overheads which are included as preliminary items in the bill of quantities, i.e. the cost of setting up and operating the site including temporary site buildings, electricity, water and telephone supply, and site security. It should be borne in mind that preliminaries are a capital cost to the contract and may not be increased if indemnifiable damage under a policy occurs, unless there is an extension of the contract period.

Another type of overhead arises in the form of administration in placing and processing orders for replacement materials and is added to the invoice value.

As regards wages, there are overheads or allowances for building projects which add to the amount of wages paid to workmen and the increased charges are about 33.5% for craft operatives and 35.6% for labourers according to Spon's *Prices for Measured Work – Major Works.* Thus from 29 June 1992 guaranteed minimum weekly earnings in the London area for craft operatives and labourers are £159.71 and £136.11 respectively. But to these rates have been added allowances for the items below in accordance with the recommended procedure of the Institute of Building in its *Code of Estimating Practice.* The resultant hourly rates on which the *Prices for Measured Work* has generally been based are £5.47 and £4.73 for craft operatives and labourers respectively.

The items referred to above for which allowances have been made are:

- Lost time
- Extra payments under National Working Rules 3, 17, and 18 (which deal respectively with continuous extra skill and responsibility, work in difficult conditions, and tool and clothing allowances.
- Construction Industry Training Board Levy
- Holidays with pay
- Accidental injury, retirement and death benefits scheme
- Sick pay
- National Insurance
- Severance pay and sundry costs
- Employers' liability and third party insurance.

There are similar guidelines in the civil engineering industry, which provide for statutory and other charges. The percentage addition is also about 35%.

Claims submitted to insurers will usually include profit, and, provided the insured contractor can show that had his resources not been employed on the repair and replacement work they would have been employed elsewhere, this profit is justified. The cost of preparing and presenting the claim is not a cost recoverable under the policy.

7.4 Clause 22 and the contractors' all risks policy

Summary of risks remaining uninsured

Only the main risks can be indicated under this heading so the list is not exhaustive. As at the end of Chapter 6, it is necessary to issue a warning that the remarks under this heading are made on the assumption that the basic policy cover indicated in this chapter is to be considered and not the cover an insurance broker by negotiation with an insurer could produce for one particular insured.

The contractor under clause 22A

Sum insured for works, unfixed materials and goods

This sum must be 'for the full reinstatement value thereof (plus the percentage, if any, to cover professional fees stated in the Appendix)'. This includes all work executed and all unfixed materials and goods. See this chapter under the heading 'The operative clause'. It is a question of choosing an adequate method of catering for inflation. While the policy should do this by an escalation clause (see earlier), the cover does not include consequential loss (see Chapter 11).

Temporary buildings, plant, tools and equipment owned or hired by the contractor

The contract does not require these items to be insured but it would be imprudent for the contractor not to do so. In fact a contractors' all risks policy could cover these items, and it is for the contractor to see that the sums insured mentioned in the schedule of the policy for these items are *adequate*. This is usually done by completion of a proposal form question. However, the contractor may well have an annual policy covering these items anywhere in the United Kingdom, at least for fire, etc. risks, and be satisfied with this.

The gap between the CAR and the public liability policy

Assuming the CAR policy cover comes to an end when the contract is completed (even if it does not, note the next heading), attention is drawn to the gap between the CAR policy and the public liability policy mentioned in the summary at the end of Chapter 6. Future structural damage (i.e. latent defects) is not usually insurable for the builder, particularly in the case of dwellings. There is a decennial insurance (covering major structural damage for ten years) available for commercially let dwellings, but this is only for owners although they can pay extra premium for the insurers to waive their subrogation rights against the builder (see below under 'The employer under clauses 22B and 22C').

Faulty design, workmanship and materials

The contract makes an exception in the insurance clause 22.2 for defective design, defective workmanship, or defective materials, so the CAR policy will not cover such risks. It is not often that a contractor is responsible for design (or at least not under the standard form), but, if it does occur, some insurers (as indicated in this chapter when dealing with this exception to the CAR policy) will cover the consequences of defective design and only exclude the defectively designed part. Furthermore, if the contractor has a design department, it is possible to cover the activities of the members of that department under a professional negligence policy. A further discussion of this aspect appears in Chapter 17 concerning the Standard Form of Building Contract with Contractor's Design, 1981 Edition. The consequences of faulty workmanship and defective materials are covered so far as they cause damage to other property covered by the CAR policy. However, the cost of repairing the faulty workmanship and/or replacing defective materials is not. Thus the producer of such work cannot get insurance cover. Product guarantee cover is not available to the builder, or is too dear.

The employer under clauses 22B and 22C

Sum insured for works, unfixed materials and goods

The same remarks apply as for the contractor insuring. The CAR policy does not cover consequential losses and the JCT contract does not refer to the insurance of such losses (except in clause 22D), but it is the employer who is likely to suffer the greatest loss in this respect. Delay in completion of the works resulting in loss of anticipated income is probably the most likely source of and resulting consequential loss. Such cover can be provided by an advance profits policy. However, there are other consequential losses from delayed completion (see this chapter); such items can be insured if they arise from physical loss or damage to the contract works. All such items are also recoverable if a successful claim can be made against the architect which would come under his professional indemnity policy, but they only add to the possibility of exceeding the limit of indemnity under that policy ('Faulty design' below).

The gap between the CAR and public liability policy

The situation is the same as the contractor insuring, but it is possible for an employer to obtain decennial insurance cover (as explained earlier), at least for commercial buildings and even possibly for blocks of flats but not otherwise for dwellings.

Faulty design

If the architect's design is found to be the cause of the Damage, the employer should be able to recover his expenses in this respect from the architect. However, there may be limits to the architect's responsibility under his agreement with the employer or under his professional indemnity policy (it is usually an aggregate limit), and his financial resources may be small. The defence of the 'state of the art' is another factor to be reckoned with in recovering from the architect, bearing in mind that it has to be proved that the architect had not reached the recognised professional standard.

Chapter 8

Project insurance

In the case of large construction projects the owner who commissions the work can waste money as a result of the system of making every participant arrange separate insurance policies for his part of the work, his own plant and equipment, and for the liabilities to workmen and other third parties that arise from performance of his part of the work.

Comprehensive project (sometimes referred to as 'wrap-up' or 'omnibus') insurance preferably arranged by the owner, has been advocated. However, when this is done, it is usually confined to conventional risks only, i.e. loss or damage to the works and to other property at the site, and to liability to third parties, including cross-liabilities between the parties involved in the contract.

Theoretically there is no reason why the architect's or consulting engineer's professional liability or that of any other professional man involved in the contract should not be included in this project insurance. The reason why this is not done as a general practice is that the professional indemnity insurance market is a small one, and as the majority of insurers do not transact this class of business, they would not be prepared to give professional indemnity cover to the professional men involved in a construction project.

A few insurers now offer to go further to help overcome the problems arising and the disputes that frequently occur between contractors and the professional men involved in the contract with regard to liability and responsibility for accidents arising from defects. They are prepared to widen their CAR policy in the first place to include:

- the owner (employer) and all the contractors he employs directly;
- the main contractor;
- all sub-contractors;
- architects, consulting engineers and quantity surveyors.

Secondly, the policy provides design damage cover so that the professional men in the last point above are covered not only for their site activities but also for their office work in connection with the project. This cuts across the cover provided by these individuals' professional indemnity policies, but it only does so far as such negligence:

- causes damage to the works (not defects without such damage);
- arises during the construction period (not the maintenance period).

This cover helps to overcome arguments as to who is to blame in tort or contract, and the resultant delays and extra costs that so easily occur, bearing in mind that the interests of all the parties mentioned in the list above are jointly covered.

It is important that the terms and conditions of any type of project insurance are disclosed to bidders in the tender documents so that all bidders will be able to cost their tenders accordingly, not forgetting to omit the contract concerned from any returns they make to their own insurers of their annual policies in order for the latter to calculate the premiums on those annual policies.

Even though the professional indemnity aspect of project insurance is omitted, the arguments are not all in favour of this type of insurance, and the following advantages and disadvantages have to be considered.

8.1 The advantages and disadvantages of project insurance

A discussion of the advantages and disadvantages of project insurance tends to become one of employer-arranged against main-contractor-arranged insurances. This is for the simple reason that the employer will include all contractors he directly employs, as well as the main contractor and all sub-contractors, in the policy he arranges for the project, whereas the main contractor will not include all these parties in his policy. However, before listing the points for and against project insurance, it must be made clear what cover is being discussed.

In project insurance the professional indemnity cover must be ignored for the reason mentioned earlier (limited cover) plus the fact that professional indemnity policies are usually arranged on a claims-made basis, thus cover virtually ceases when the policy lapses (i.e. when the contract is completed). While the CAR policy with a public liability extension will also often cease on completion of the contract plus a limited form of cover during the maintenance period; this policy is on a losses-occurring basis and consequently the cover is wider. Professional indemnity claims particularly in the construction industry, 'have a long tail', i.e. defects can arise over many years after the contract is completed.

In project insurance it is normally only the CAR and public liability risks that are included. This leaves as a residue the employers' liability and motor insurances outside the project cover. As these two are legally compulsory in the UK, it is probably better to leave such insurances for the participants in the project to make their own arrangements.

Disadvantages of project insurance

Contractors will give the following drawbacks in employer-arranged insurances:

(1) The employer's insurers cannot know, at the time when the covers are being arranged, even the identity of the contractor who is going to carry out the work and certainly nothing about the efficiency of the contractor's organisation and his insurance claims record. Therefore it is difficult to see upon what basis the risks insured can be rated. Consequently, most underwriters would load premium rates and excesses of CAR insurance arranged by the employer, where the identity of the successful contractor is unknown.

(2) If, because of the foregoing, the employer's insurance is not arranged until after the contract is let, then there are going to be delays in effecting cover which do not occur where the contractor is responsible for his own insurances. This is because the contractor will almost invariably (except for overseas projects) have an annual policy on a post-declaration basis so that cover is always in force in respect of any contract undertaken.

(3) Contractors carry out many contracts and therefore are in the best position to know the risks against which they should insure. The contractor's policy will be tailored to his needs and the activities which he undertakes (see, for example, the extensions available to the CAR policy detailed in Chapter 7). Furthermore, the contractor is constantly in the market and dependent on his claims experience. Thus he is in a better position to obtain competitive rates in respect of the risks which are insured.

The employer, on the other hand, will frequently only be undertaking a once-in-a-lifetime construction project and will lack the 'muscle' in the insurance market quite apart from the lack of knowledge of where to go for specialist and best advice. In any event, it is doubtful whether the employer will have sufficient detailed knowledge of such factors as the temporary works, as these will only become apparent when they have been designed by the contractor (e.g. cofferdams and falsework to bridges, etc. are usually the contractor's responsibility): according to the *Bragg Report on Falsework*, temporary works failures have been the cause of large insurance claims.

(4) Frequently there are residual risks which are stated not to be included in the covers effected by the employer. These have to be insured by the contractor, usually under his annual policy, and the resultant saving by the contractor between the full risks and the residual risks can be marginal. In fact it has been known for employer-arranged CAR cover purporting to be project insurance not to cover the temporary works of the constructional plant, temporary buildings and other property of the contractor.

Although, on the face of it, this may not seem unreasonable as he owns most of these items, if the contractor has to insure this property separately, the cost can be prohibitive as it is the hazardous part of the risk. Similarly, some insurers are unable to add a public liability section to the CAR policy, particularly on overseas projects, as their reinsurance facilities do not allow this and again the premium cost would be heavier. Furthermore the overall premium (employer's and contractor's) is higher than if one policy handled all the risks.

(5) There is an argument for saying that as the contractor is usually responsible for the works under the contract, he should also be empowered to take steps to protect himself, including effecting insurance, whereas when the employer seeks to arrange the contract insurances he still leaves responsibility for the works with the contractor. Similarly, if the employer insures the project, the contractor should be relieved of responsibility for any inadequacies in the insurance cover to the extent that it falls short of the contract requirements.

(6) A contractor usually has a long-standing relationship with his insurers and the loss adjusters his insurers use, and claims are expeditiously and often generously settled. Therefore a contractor does not take kindly to an employer-arranged package which falls short of those arranged by the contractor. Consequently, it will often be necessary for the contractor to arrange a 'difference in conditions' policy in order to avoid exposure to any risk attaching to him and normally covered under his own insurances. For example, the JCT form and the ICE Conditions require insurances for off-site goods certified and paid for, but it is doubtful whether an employer-arranged CAR cover would always cater for these risks or for the transit cover of such goods. As mentioned in (4) above, separately arranged off-site goods and transit cover can be expensive for the contractor.

Advantages of project insurance

In spite of what is said by contractors, certain advantages do exist for the employer-arranged project insurance.

(1) The very strong argument is that mentioned at the beginning of this consideration of project insurance, namely, it avoids the time and trouble spent by all parties involved in a loss, damage or liability claim in blaming the other parties to the contract or sub-contract.

(2) In very large projects, particularly where civil or mechanical engineering as well as building work is involved, with many sub-contractors being used, project insurance covering all interested parties avoids gaps in the individual parties' cover.

(3) The employer, in one way or another, pays the premium, even when the contractor arranges insurance cover. Therefore the employer has the right to protect himself by buying project insurance. The contractor regards this as a theoretical argument as the employer is not often in a position to provide more adequate insurance than the contractor, or at least to match his cover. However, assuming the employer is properly advised, it could apply, particularly in the case of the inadequately insured sub-contractor. In any event, the cost of the insurance to the employer can only be defined in any detail

when the main contractor is known. The cost can be only broadly established before the contractor is selected.

(4) Subject to the employer obtaining the best advice:

- The use of excesses can be controlled.
- Loss or damage by design failures can be minimised as the employer would control the overall cover; and the settlement of claims would be quicker and easier as it would be a matter of negotiation between the employer's brokers and the insurers or their loss adjusters (with the consequent improvement in overhead costs and cash flow on insurance claims).
- Use and occupancy problems can be catered for easily.

8.2 UK constructions involving project insurance

The Thames Barrier Scheme, details of which are set out below, and the Eurotunnel Project in Appendix 7 provide examples of project insurance. The Thames Barrier Scheme was carried out under a specially arranged CAR policy, and incidentally that scheme illustrates what could be achieved by project (package) insurance for a very large contract even as long ago as the early 1970s.

The CAR insurance was placed with a leading office and eight other UK insurers plus Lloyd's. The policy covered the contract works, temporary works, materials and non-marine construction plant (marine hull insurance had to be effected elsewhere) tools and equipment, together with a public liability section. The main points of this insurance were:

- All sub-contractors covered as joint insureds
- Cover for all operations anywhere in the UK including storage and transit risks whether by land or water
- Indemnity for loss or damage caused by defective design, materials or workmanship *without* exclusion of the defective part
- Indemnity for increased cost of working sustained in expediting the reinstatement of loss or damage to the insured property in order to avoid or reduce a delay in completing the contract, subject to a limit of £2 million
- Cover for the cost of professional fees, removal of debris, shoring up etc. without separate lower limits
- Cover for permanent buildings on the site used by the contractors in relation to the contract
- Deductibles (excesses) of:
    (i)   £25 000 each claim for defective design, materials or workmanship,
    (ii) £10 000 each claim for property in the river or dock, and
    (iii) £500 each claim for all others

Currently these excesses would be very much higher.

Bearing in mind that the insurance market when this contract was placed was less competitive than it is nowadays, it is felt that at the time of writing it would be easier to place the insurances with possibly lower rates and wider coverage. For example, it is possible that a separate marine hull insurance could be arranged in one omnibus policy with the CAR, etc. cover.

It is interesting to note that in this contract there were problems with ground conditions but the costs of overcoming these were paid for under the equivalent clause to clause 12 of the ICE Conditions of Contract dealing with adverse physical conditions and artificial obstructions which could not reasonably have been foreseen by an experienced contractor.

Chapter 9

JCT 80 clauses 21.2.2 to 21.2.4

While basically this sub-clause 21.2.1 calls for a liability insurance policy, it is a hybrid in the sense that it covers the insured's (the employer's) property, although not the works, just as much as it does his legal liability for damages to third party property.

9.1 Detailed consideration of clause 21.2.1

Clause 21.2.1 under JCT 80, 1986 version, reads as follows:

21.2.1   Where it is stated in the Appendix that the insurance to which clause 21.2.1 refers may be required by the Employer the Contractor shall, if so instructed by the Architect, take out and maintain a Joint Names Policy for such amount of indemnity as is stated in the Appendix in respect of any expense, liability, loss, claim or proceedings which the Employer may incur or sustain by reason of injury or damage to any property other than the Works and Site Materials caused by collapse, subsidence, heave, vibration, weakening or removal of support or lowering of ground water arising out of or in the course of or by reason of the carrying out of the Works excepting injury or damage:
  .1.1 for which the Contractor is liable under clause 20.1;
  .1.2 attributable to errors or omissions in the designing of the Works;
  .1.3 which can reasonably be foreseen to be inevitable having regard to the nature of the work to be executed or the manner of its execution;
  .1.4 which is the responsibility of the Employer to insure under clause 22C.1 (if applicable);
  .1.5 arising from war risks or the Excepted Risks.

Sub-clause 21.2.1 replaces the previous clause 19(2)(a) under the JCT 1963 Edition which was introduced following the decision in *Gold* v. *Patman and Fotheringham Ltd* (1958). In that case the contract was the 1939 Standard Form of Building Contract, 1952 revision, where clause 15 required the contractors, Patman and Fotheringham, to 'effect ... such insurances ... as may be specifically required by Bills of Quantities'. The risks required to be insured by the bills included 'insurance of adjoining properties against subsidence or collapse'. A neighbouring owner brought a claim against the employer, Gold, in respect of subsidence caused by the building operations. The contractors had taken out a policy covering their own liability for subsidence, but not that of the plaintiff. The contractors had not been negligent (hence there was no right of action against the contractors by the neighbour) but Gold was liable for removing the support to his neighbour's land (however this was done). He tried to recover from the contractors but the Court of Appeal held that they were not in breach of their obligations under the contract. It was clear from this decision that employers could not rely on the liability and indemnity clause of the JCT contract (now clause 20) for their complete protection where a claim was made in these circumstances. The Court of Appeal regarded the contract shown by the specific provision for fire insurance (now clause 22) being in the joint names of the employer and contractor, and the requirement for the policies now referred to in clause 21.1 as significant. Thus, where the contract requires the employer to be insured by the contractor it expressly so provides.

It was decided to introduce a clause into the contract whereby insurance would be required to protect the employer for losses caused otherwise than by negligence of the contractor. However, the first wording used in 1963 was unsatisfactory as it did not set out the perils which were to be covered, nor did it exclude the works or mention exceptions. The amendment in 1968 listed the perils mentioned above, excluded the works and listed exceptions. Nevertheless, there was still some doubt as to exactly how the insurance was to be requested, i.e. whether a provisional sum specifying limits of indemnity in the bills was sufficient or a specific instruction by the architect was required.

Arranging the insurance

The introductory wording in the current clause of the 1986 amendment reads:

> Where it is stated in the Appendix that the insurance to which clause 21.2.1 refers may be required by the Employer the Contractor shall, if so instructed by the Architect, take out and maintain a Joint Names Policy...

The doubt as to whether a provisional sum in the bills was sufficient to implement this insurance, or whether instructions were also required from the architect, was resolved by the 1986 amendment. The correct procedure is now stated in the introduction to the clause. Both a statement in the Appendix that this insurance may be required by the employer and also an instruction by the architect are necessary. The requirement for a provisional sum to be included in the contract bills has been omitted.

The clause continues as follows:

> ... for such amount of indemnity as is stated in the Appendix in respect of any expense, liability, loss, claim or proceedings which the Employer may incur or sustain...

An entry in the Appendix is provided for the amount of indemnity which may be for any one accident or series of accidents arising out of one event or an aggregate amount.

The architect should consult the employer before work is begun as to the amount of indemnity required, bearing in mind that the risks referred to in clause 21.2 arise mainly when the works are adjacent to existing buildings or main services, and it is the value of these that is at risk and covered by this insurance. The object is for the contractor to obtain a quotation for the employer for the issue of a policy in the joint names of the employer and contractor which the architect, with the employer's approval, can instruct the contractor to accept. Probably the quickest way for this to be done is for the architect, contractor and insurance official to meet on the site so that the risks can be examined and ideally a quotation obtained and accepted on that occasion, assuming the architect and insurance official have the necessary authority.

It is clear that the contractor is only an agent required to arrange the insurance to indemnify the employer, and although the contractor is a joint insured, he is not required to indemnify the employer; also the second part of this sub-clause refers to expense, etc., which the employer may incur. The last two points support the statement that the only undertaking by the contractor is to arrange the insurance. The reasons submitted for this stipulation for joint names are that it is a protection for the employer and a satisfaction to the insurer. This is because there is a contractual relationship between the contractor and the insurer involving direct responsibility for the premium and for the information disclosed on the proposal form which is incorporated into the policy. Both of these direct responsibilities of the contractor would not exist if he could only be regarded as an agent and not as a joint insured. Usually the contractor is well known to the insurer as the insurance is often placed with the contractor's public liability insurer. It is more satisfactory for the insurer to enter into a contract with a party whose financial standing and honesty are to some extent known than to contract only with the employer, who is usually unknown to the insurer.

Furthermore, if the insurer decides to stipulate for the work to be carried out in a certain way, he will have more control over the contractor who is a joint insured. However, the fact that the contractor cannot claim under the policy can cause misunderstandings in the handling of claims.

The phrase 'expense, liability, loss, claim or proceedings' is wide enough to include consequential loss, particularly as elsewhere in the contract (e.g. clause 22.2) the words 'physical loss or damage' only are used where clearly only insurance cover for physical damage is required.

Subsidence, etc.

The clause continues:

> ... by reason of injury or damage to any property other than the Works and Site Materials caused by collapse, subsidence, heave, vibration, weakening or removal

of support or lowering of ground water arising out of or in the course of or by reason of the carrying out of the Works.

As the works are excluded, the property concerned can only belong to third parties or to the employer, which is extraneous to the works, but it does not have to adjoin the works, although it usually does.

'Heave' has now been added to the list of causes of the damage and it may occur in clay soil if the moisture content is increased; this can arise as a result of removal of tree roots or vegetation which have previously been taking water from the soil. Heave could give rise to claims under the 21.2.1 type of policy on, say, a housing estate where houses have been sold or handed over and are thus no longer part of the works.

The perils of collapse, subsidence, heave, vibration, weakening or removal of support are all risks which insurers catering for the building trade are prepared to cover. In *David Allen & Sons (Bill posting) Ltd* v. *Drysdale* (1939) Lewis J decided that

'"Subsidence" means sinking, that is to say, movement in a vertical direction as opposed to "settlement", which means movement in a lateral direction, but I am of the opinion that the word "subsidence" in this policy covers subsidence in the sense in which I have defined it, and also settlement.'

In the same case it was said 'Collapse denotes falling, shrinking together, breaking down or giving way through external pressure, or loss of rigidity or support.' Unless a policy excludes 'settlement' it is probably inadvisable to attempt to do so by a definition which comes from outside the policy.

The effect of ground water

The lowering of ground water levels is a circumstance so unfamiliar to the average insurer in that it is not a wording normally used in a subsidence exclusion. However, it can obviously have the same effect on property as vibration or the removal or weakening of support.

Below the surface of the ground, at varying depths, is a large sheet of water termed the ground water. Should the level of the ground water be rather high, work in the ground may involve a permanent or temporary lowering of the ground water by draining of the sub-soil. The lowering of ground water can occur as a result of the following activities, among others:

- The construction of sub-soil drains which provide a remedy for dampness caused by waterlogged soil.
- Certain types of piling involving the removal of ground water.
- Pumping from open sumps, or by well points, when constructing below the ground water table, until such construction is completed.

The removal of ground water, if substantial or over a long period, can cause subsidence. If mud or soil in suspension appears in the water being extracted or diverted, the builder should take steps to hold back the soil being carried with the water, because this is erosion which is a first step to subsidence.

Natural right to support for land

At law every *landowner as distinct from the owner of a building* has a natural right of support, which means a right not to have that support removed by his neighbour. However, there is no natural right at law to support for buildings. 'The owner of the adjacent soil may with perfect legality dig that soil away, and allow his neighbour's house, if supported by it, to fall in ruins to the ground.' See *Dalton* v. *Angus & Co* (1881). However, if in these circumstances, the land on which the buildings stood would have fallen whether built upon or not, then an action is allowed in respect of damage to the buildings. Assuming damage to the land would have been negligible if no building had been there, the building owner's only right of action is to sue in negligence, i.e. he has no automatic right of action which the pure land-owner (without buildings) has.

Right of support to buildings

This absence of a natural right in the case of the building owner applies similarly to the support provided by one building to another. Nevertheless, the right to have buildings supported by land or by other buildings can be acquired as an easement, which does not automatically accompany own-ership but must be acquired by grant, either express, implied or presumed.

- *By express grant or reservation.* Here the deed creating the grant will specify its exact nature. On the sale of a building, for example, the owner may expressly grant an easement of support in favour of the grantee over the land or building retained, or he may expressly reserve for himself a right of support over the building sold.
- *By implied grant or reservation.* In these cases there is a mutual easement of support between two adjacent buildings where the construction is such that each building must give some support to the other, e.g. semi-detached houses and terraced houses.
- *By presumed grant.* Where a right to support has been enjoyed for a long time, it is the law that it shall not be defeated merely because the person exercising the right cannot produce the grant concerned. In these cases the law presumes that a valid grant was made. The acquisition of easements in this way is known as 'prescription' and is governed mainly by the Pre-scription Act 1832. As a generalisation this provides that where a building has benefited from the support of another for the period stated (normally twenty years) without interruption, then the building will acquire a right to continued support.

Consequently in *Dalton* v. *Angus & Co* (1881), where one of two adjoining houses was converted into a coach factory which caused more pressure upon the other house and was so used for over 20 years, the House of Lords held that an action lay for demolishing the other house and so causing a part of the factory to collapse.

The natural right protected by tort

A natural right which exists automatically is protected by the law of tort. Thus the removal of support to land (as distinct from buildings) constitutes a nuisance which is actionable. Generally speaking, in these cases there is no necessity to consider the circumstances leading to the removal of the support, as is essential in the case of negligence, because the liability arises from the neighbour's decision to carry out work which creates a risk of removing support from the adjoining property. However, there is an exception because if, by drainage on his own land of water percolating from beneath neighbouring lands, an occupier draws off all such underground water, he becomes the owner of it and will not be held liable to neighbours though the effect may be to dry up the neighbouring land and remove support. Statute may, however, affect this exception.

In *Langbrook Properties Ltd* v. *Surrey County Council* (1969) an example occurred of an occupier draining on his own land water percolating beneath neighbouring land without liability to the neighbour for the resultant dewatering and subsequent settlement of buildings. The plaintiff company, owners of a site which it was developing, claimed damages for nuisance and negligence against the defendants, alleging that by pumping out excavations on land adjacent to the plaintiff's land during the first six months of 1968, the defendants had abstracted water percolating beneath the plaintiff's land, causing dewatering, with the result that differential settlement of land and buildings took place. The defendants maintained, *inter alia*, that the statement of claim disclosed no cause of action. A preliminary issue was ordered to be tried to determine whether the plaintiff had any cause of action against the defendants or some one or more of them by reason of withdrawal of water by means of pumping from beneath the surface of the first defendant's land.

Plowman J said that the question is whether a man whose land has subsided as a result of the abstraction by his neighbour of water percolating under that land can in law maintain an action for consequential damage either in nuisance or in negligence. The action was not about water flowing in a defined channel, nor was it concerned with easements, either of support or of any other nature; nor did any question of derogation from any grant arise. The authorities established that a person might abstract water under his land which percolated in undefined channels to whatever extent he pleased, notwithstanding that that might result in the abstraction of water percolating under the land of a neighbour and thereby cause him injury. In such circumstances there was damage but no violation of legal right, and thus no legal remedy. There was, then, no room for the law of nuisance or negligence to operate. If there had been, it was highly probable that the courts would have already said so. In *Chasemore* v. *Richards* (1859) the opportunity was there, since the water authority concerned was found to have had reasonable means of knowing the natural and probable consequences of their excavations, but there was no suggestion in the House of Lords that this was a relevant matter. Moreover, since it was not actionable to cause damage by the abstraction of underground water, even where that was done maliciously, as in *Bradford Corporation* v. *Pickles* (1895), it would seem illogical that it should be actionable if it were done carelessly. A claim in nuisance could fare no better. Nuisance involved an unlawful interference with a man's enjoyment of land, and the interference, as the authorities showed, was not unlawful. The question posed in the order had to be answered in the negative.

It was stated earlier that if mud or soil in suspension appears in the water being extracted or diverted, the builder should take steps to hold back the soil being carried with the water because this is erosion, which is a first step to subsidence. Now the important question in the light of the *Bradford Corporation* and *Langbrook Properties* cases is whether the defendants would still be free from liability for such *erosion* and resultant subsidence by abstracting percolating water. The law report in the *Langbrook Properties* case gives no indication whether erosion took place or not. Undoubtedly, as stated, there was dewatering causing subsidence. If, for example, the percolating of the water beneath the plaintiff's land had been speeded up due to the pumping out of the excavations on the land adjacent to the plaintiff's land to such an extent that erosion took place, would there be any liability? The basic principle of law is that no interest in percolating water exists until appropriation, therefore no interest or right can be infringed. However, if soil as well as water is extracted, it is clearly arguable that a right to support of land is affected by more than the removal of percolating water. The court could therefore come to a different decision. The difficulty would be proving the extraction of soil and that it caused the subsidence or at least contributed to it as much as the dewatering. The expense of experts in these cases is discouraging to insurers who would wish to know the law, but sooner or later a case will arise which economically justifies the expense.

The **Redland Bricks** *case*

The House of Lords in *Redland Bricks Ltd* v. *Morris and Another* (1969) considered the remedies available where excavators had removed support to neighbouring land. The respondents were market gardeners who farmed eight acres of land; this was adjoined by the appellant's land which the appellant used to dig for clay. In 1964 some of the respondents' land slipped, due to lack of support, into the appellant's land. Slip occurred again in 1965 and 1966. It was likely that further slips would occur rendering a large part of the respondents' land unworkable as a market garden. To remedy the slipping was estimated to cost about £30,000. The respondents' land was worth £12,000. In October 1966 a judge (upheld by the Court of Appeal) granted two injunctions in favour of the respondents:

- an injunction restraining the appellants from withdrawing support, and
- a mandatory injunction 'that the [appellants'] do take all necessary steps to restore the support to the [respondents'] land within a period of six months.'

On appeal against the mandatory injunction, it was held that, although there was a strong probability that grave damage would, in the future, accrue to the respondents, the injunction would be discharged because in its terms it did not inform the appellant exactly what it had to do. As the appellant had behaved, although wrongly, not unreasonably, it would have been wrong to have imposed on it the obligation of remedying the slip at a cost of £30,000, which would have been unreasonably expensive. The judge would, however, have been justified in imposing an obligation to do some reasonable and not too expensive works which might have had a fair chance of preventing further damage. Thus the decision of the Court of Appeal was reversed on this point.

These legal aspects must be considered by insurers in the event of a claim under any policy they issue covering the perils mentioned in sub-clause 21.2.1 of the JCT contract.

Exceptions

It will be seen that the exceptions of clause 21.2.1 which now follow form the basis of exception 1 of the clause 21.2.1 policy which is considered later in the next chapter.

Clause 20.2 liability

The clause continues:

> excepting injury or damage:
> .1.1 for which the Contractor is liable under clause 20.2;

The purpose of this exclusion is to ensure that the negligence, etc. of the contractor, etc. is not covered by this insurance, as clause 21 (requiring insurances to cover the liability imposed by clause 20) deals with this situation. Incidentally, if the public liability policy of the contractor were arranged in the joint names of the contractor and employer the latter would be covered and possibly (if the policy was wide enough), a 21.2.1 cover would not be necessary. However, such a public liability policy would not in any circumstances cover property belonging to the insured, whereas the 21.2.1 policy does cover property belonging to the employer other than the works.

standard public liability policy would cover these circumstances, because they are 'accidental' as far as the insured contractor is concerned, unless caused by third parties. Nevertheless, it should be mentioned, as experienced underwriters will know that circumstances not envisaged at the time of drafting a policy do arise.

Consider the position of the employer's supplier's lorry driver who dumps a load of paving stones and heavy wet sand against some underpinning of the first floor of the works, which inevitably causes a collapse affecting third party property that has an implied or presumed right of support. Assuming the wrongdoer is uninsured and a man of straw, the employer will invoke his 21.2.1 cover because sub-clause 20.2 and the complementary insurance sub-clause 21.1.1 will not help as there is no negligence of the contractor or of those for whom he is responsible. The only possible exception to sub-clause 21.2.1 which could apply is exception 3. However, this type of inevitability clearly is not reasonably foreseeable having regard to the nature of the work to be executed or the manner of its execution, i.e. it does not arise from performance of the work but from an extraneous or casual act.

Employer's responsibility to insure

There are further exceptions to injury or damage:

> .1.4 which is the responsibility of the Employer to insure under clause 22.C.1 (if applicable);

In the 1986 amendment the drafters have at last appreciated that this exclusion in the 1980 edition, which read 'which is at the sole risk of the Employer under clause 22B or 22C (if applicable)', was largely unnecessary as the works are already excluded and clause 22 is generally speaking limited to the works (using this term as including site materials) although it has to be admitted that existing structures in clause 22C justify the exclusion of existing structures. This is now made clear by confining the exclusion to 22.C.1 which concerns existing structures and the contents thereof owned by the employer or for which he is responsible. Furthermore the basic sub-clause 21.2.1 now excludes 'Works and Site Materials'.

War risks and excepted risks

> excepting injury or damage:
> 1.5 arising from war risks or the Excepted Risks.

War risks have been considered in Chapter 6. Excepted risks are defined in clause 1.3 as:

> ionising radiations or contamination by radioactivity from any nuclear fuel or from any nuclear waste from the combustion of nuclear fuel, radioactive toxic explosive or other hazardous properties of any explosive nuclear assembly or nuclear

component thereof, pressure waves caused by aircraft or other aerial devices travelling at sonic or supersonic speeds.

Nuclear risks and sonic waves were also dealt with in Chapter 6.

9.2 *General matters concerning clause 21.2.1*

Having considered the meaning of sub-clause 21.2.1 in this chapter it is relevant to make the following points.

Cover for the employer

All individual firms and companies having building work done, whether under the JCT Standard Form of Contract or not, should ensure that their own insurances cover their liability for this work; and in particular that any risks or liabilities for which they are not obtaining an indemnity are covered as far as possible, even if it leads to dual insurance. Unfortunately Hodson LJ in *Gold* v. *Patman* gave the impression that the contractor should be the agent in effecting any additional insurances required by the employer, when he stated:

> 'It is plain from the nature of the contract that there would be no objection to a provision for the insuring of the building owner by the contractors being included in the bill of quantities as an obligation of the contractors . . .'

Thus the tendency is for the employer to turn to the contractor instead of making his own arrangements. Worse still is the tendency for those seeking such insurance to attempt to widen sub-clause 21.2.1 to provide an employers' liability and public liability cover for the building activities, which should be arranged directly for the employer without involving the contractor. For example, a biscuit manufacturer who is about to have another factory built will have both an employers' and a public liability policy but probably not covering building work. Those policies will have to be extended to cover this work, but this can be done without involving the contractor. From the insurer's viewpoint the overlapping of the 21.2.1 policy with other policies which might be giving the employer cover is inadvisable, especially when the insurers are different. These aspects, i.e. the direct arrangement of insurances and the avoidance of overlapping, are developed in the next chapter, and specific cover for the employer is now considered.

Extending existing cover

The types of principal, person or firm having building work done will vary tremendously and each will have a range of existing insurances. Some of these contracts will be evidenced by standard forms of policy and others may be specially tailored to suit the insured's requirements. However, all will be arranged to cover the insured's activities in a certain occupation (usually not building work). Therefore it is extremely unlikely that the existing insurances held by the employer in the building contract will be adequate to cover a completely different activity from his normal one, namely the building work proposed.

It is true that the employer will obtain indemnities from the contractor under the Standard Form of Contract or any other standard form which may be used, but it will be appreciated from what has been said that these indemnities are never complete. Secondly, there is always the possibility, however slight, that the contractor fails to implement the insurances required under the contract or they fail to operate for some reason outside the employer's control, and a contractor's liability without adequate insurance backing may not be much use.

Therefore two basic steps are necessary. The first and most important is for the employer to warn his existing insurers that he is having building work done. Secondly, he should, through his agents, ensure, as required by the Standard Form of Contract, that the contractor has taken out the insurances required of him under the contract. The employer should at the same time consider his own additional insurance requirements. A supplementary policy with 21.2.1 cover as a basis has been mentioned but an alternative is for the insurance industry to provide protection without any 21.2.1 basis. Probably the correct approach is to consider this cover under the three headings, material damage, consequential loss and liability.

Material damage

Taking the material damage heading first, it seems that an all risks policy is the most obvious cover because it is the widest available, and, assuming the contractor is required to insure the works on an all risks basis, there are still other structures to consider. The following aspects must be resolved with regard to existing structures which are not part of the works, and which belong to the employer, although if the employer is responsible for the works etc. under clauses 22B or 22C then his existing policies (commercial all risks or fire and special perils) could be made to apply.

(1) For his own existing structures the employer will probably have a fire and special perils policy.

(2) Alternatively he will have the modern cover for industrial buildings which is the 'all risks' policy against accidental loss or destruction of or damage to the specified property from any cause other than those excluded.

(3) In the case of both (1) and (2) the employer must bear in mind the need to transfer the new structures to his own insurances or make alternative arrangements at the stages of sectional or practical completion. However, in the case of the commercial all risks policy a standard exclusion is 'property in the course of erection, alteration or demolition or whilst

under process'; therefore this exclusion will have to be deleted if the building during course of erection is to be covered. In any event insurers will have to be informed if either of the policies in (1) or (2) are to provide any protection during the actual erection.

(4) It has already been mentioned and will be mentioned in more detail under the heading 'The limitations of the policy' below, that the clause 21.2.1 policy gives the employer some protection for his property in the vicinity of the works, but the perils are limited. So also are the perils in the fire and special perils policy. Therefore the 'all risks' cover is clearly the best, at least for existing structures in the vicinity of the works.

(5) Property developers selling or letting new buildings may consider decennial insurance for structural defects, i.e. latent defects, which is now available in a very limited market in the UK. This is mainly for commercial buildings and possibly blocks of flats. The National House-Building Council's scheme provides a ten-year cover to purchasers of private dwellings (and their successors in title) for structural defects.

Consequential loss

Regarding consequential loss the modern 'all risks' form of contract insures not only property but also earnings (i.e. the business interruption risk) in the one policy. However, if the building in course of erection is to be sold or let then delay in completion cover in the shape of advance profits insurance is necessary.

Liability

Under the heading of liability a contingency insurance may be considered necessary to pick up those liabilities (subject to the usual terms and conditions to be agreed) which the existing employers' and public liability policies of the building owner/developer may not cover, including failure of the contractor's indemnities or his insurances to operate for one reason or another leaving the employer with a legal liability. The 21.2.1 cover would have to be taken into consideration in this arrangement.

Arranging the cover

In practice, 21.2.1 cover is often arranged by the employer whose occupation is that of a property developer. If there is no Appendix statement requiring this additional insurance and the employer's payment for it, the contractor need take no action; even when there is, the employer sometimes also stipulates that a certain insurer will provide the cover and the employer's quantity surveyor or architect deals directly with the insurer. Possibly the contractor is glad to leave the employer's agent to arrange an insurance which does not benefit him and only saddles him with the job of an insurance agent.

The limitations of the policy

The 21.2.1 policy is a hybrid in that it gives liability, material damage and consequential loss insurance cover within the same operative clause. All property which is adjacent to the site is vulnerable as a result of the building activities. No doubt the employer's adjacent property will be covered by a fire and special perils policy which would not include the subsidence risks mentioned in sub-clause 21.2.1. However, the employer's property may suffer from other perils not covered by a 21.2.1 policy or a fire and special perils policy. An illustration is dust, which does not have to be excessive when arising from building work to cause serious damage to delicate machinery and electrical equipment such as computers (even though well protected) in neighbouring buildings. Thus there may be no liability on the contractor because in negligence the vital question is, 'Did the defendant take reasonable care?' If he did, there can be no liability.

Consequential loss

Whereas consequential loss following damage to the contract works is excluded, because the wording specifically excludes the works, consequential loss sustained by the employer following damage to third party property, to the employer's property or to that for which he is responsible would be covered. This can result in a number of claims which may not be foreseen. For example, delay in completion of the works resulting from damage to adjacent property in the *Gold* v. *Patman* situation can arise from the blocking of entrances to the site. It is difficult to argue that this is not an expense which the employer has incurred by reason of damage to property within the meaning of the sub-clause, because it does not arise out of damage to the works, which is excluded.

Delay in completion of the works due to interference with the contractor's work schedule by reason of damage to property (accepting the latter to be a valid 21.2.1 claim) is not so clearly an expense under the sub-clause as it could be due as much to the inability to find the necessary labour or other factors as due to the damage to adjacent property. Here it is a matter of what is the proximate cause of the delay. If the damage is the proximate cause of the delay, that delay is covered by the policy.

The situation should not be confused with the position where adjacent property collapses (say a legitimate 21.2.1 claim) on to the works, damaging them (the damage being specifically excluded under the 21.2.1 policy) and resulting in a delay-in-completion claim. This may arise because workmen otherwise used to progress the works have to clear debris or carry out repairs and the contract works schedule suffers, or they have to be paid overtime at the employer's expense in order to maintain the time schedule. In these circumstances the answer seems to lie in the basic fact that an expense which the employer may incur by reason of *damage to the works* is specifically excluded, and thus not a valid claim. It is a consequential loss following damage to the contract works. The fact that indirectly it is due to damage on adjoining property is irrelevant in the face of the specific exclusion. An advance profits policy may provide some cover for the employer.

9.3 *Approval of the insurers and costs of the insurance*

The first two sub-clauses read as follows:

21.2.2 Any such insurance as is referred to in clause 21.2.1 shall be placed with insurers to be approved by the Employer, and the Contractor shall send to the Architect for deposit with the Employer the policy or policies and the premium receipts therefor.

21.2.3 The amounts expended by the Contractor to take out and maintain the insurance referred to in clause 21.2.1 shall be added to the Contract Sum.

Sub-clause 21.2.2 is self explanatory. For sub-clause 21.2.1 insurance, the policies and receipts should be deposited with the employer before the contract starts, but as the work often begins before the insurance negotiations, it could be well advanced before the policy is deposited. By implication the final decision on adequacy of cover is that of the architect or his insurance adviser, but see sub-clause 21.2.4 below. Presumably any member of the Association of British Insurers could be approved under this sub-clause, but this is not the only test, as any statutorily authorised insurer could be approved.

Reference to Chapter 4 shows that whatever the contract clauses say about inspection of the contractor's insurances (and in this case the insurance arranged by the contractor) it often falls upon the architect to inspect the insurance for approval on behalf of the employer. Because the contract specifies exceptions in clause 21.2.1 this should make the inspection easier for the architect.

The 1980 clause 21.2.2 in the private edition required the insurance to be taken out with insurers approved by the architect, and the policy to be deposited with the architect. However, the 1986 clause 21.2.2 is similar to the 1980 local authorities edition in that the insurer is as approved by the employer. The contractor now has to send to the architect, for deposit with the employer, the policy or policies and the premium receipts instead of depositing these documents directly with the employer. Clause 21.2.3 adds the words 'take out and'; otherwise it is exactly the same as in the 1980 edition.

9.4 *Default in insuring*

21.2.4 If the Contractor defaults in taking out or in maintaining the Joint Names Policy as provided in clause 21.2.1 the Employer may himself insure against any risk in respect of which the default shall have occurred.

The words in 1980 clause 21.2.4 'and the amounts paid or payable by the

Employer in respect of premiums shall not be included in the adjustment of the Contract Sum' have been omitted from 1986 clause 21.2.4. Because insurance under 1986 clause 21.2 is no longer to be the subject of a provisional sum, it is no longer necessary to include these words.

Chapter 10

JCT 80 clause 21.2.1: insurance requirements

10.1 Introduction

Whether 21.2.1 insurance cover is required

This decision will depend upon the kind of work the contract requires. For example, piling, the removal of support to existing buildings or excavating near their foundations are clearly relevant, as is the type and condition of property adjacent to the site of the works. The greater the hazard the greater the requirement for this type of cover, i.e. there is a natural selection against the insurer. Thus a single-storey school to be erected in the middle of a large field involves little or no 21.2.1 risk. However, work on old property in the middle of a city, involving work near the foundations of third party buildings, carries a very high risk.

Limit of indemnity

The amount of the required indemnity entered in the appendix to the JCT contract can apply either to any one occurrence (or series of occurrences arising out of one event) or in the aggregate, i.e. the figure applies overall to the period of insurance. In any event the choice of amount bears no relation to the amount of the contract price. It is the value of adjoining property that has to be considered, not forgetting the contents thereof. Insurers catering for this risk can accept high limits, e.g. up to £10 million. Where higher sums are called for, co-insured or 'layered' policies would be arranged with other insurers or reinsurers.

10.2 The policy wording

Section 4 of the policy set out in Appendix 2 will be taken as an example.

The operative or insuring clause

The operative or insuring clause reads as follows:

> In the event of the Insured entering into any contract or agreement by which the

Insured is required to effect insurance under the terms of Clause 21.2.1 of the Joint Contracts Tribunal Standard Form of Building Contract (or any subsequent revision or substitution thereof) or under the terms of any other contract requiring insurance of like kind the Company will indemnify the Employer in respect of any expense liability loss claim or proceedings which the Employer may incur or sustain by reason of Damage to any property other than the Contract Works occurring during the Period of Insurance within the Territorial Limits and caused by

(a) collapse
(b) subsidence
(c) heave
(d) vibration
(e) weakening or removal of support
(f) lowering of ground water

arising out of and in the course of or by reason of the carrying out of the Contract Works

Provided that

1. the company shall not be liable for any amount exceeding the Limit of Indemnity
2. the insured shall notify the Company within 21 days of entering into or commencing work under such contract or agreement whichever is the sooner together with full details of the contract
3. once notified the Company may give 14 days notice to cancel the cover granted by this Section in respect of such contract or agreement or alternatively provided a quotation which may vary the terms of this Section
4. the indemnity provided by this Section in respect of such contract or agreement shall terminate 14 days from the date of issue of the quotation if the quotation has not by then been accepted by the Insured or the Employer.

The cover provided

This specimen policy adopts the practice of the majority of insurers and follows the wording of clause 21.2.1.

The policy then continues with the words:

by reason of Damage to any property other than the Contract Works occurring during the Period of Insurance within the Territorial Limits ...

So the policy covers damage to any property other than the works. The words 'Damage' and 'Contract Works' are defined in the policy under the 'Definitions' section. The former includes 'loss' and the latter includes 'materials supplied' and 'materials for use' by reason of or in connection with the contract. The 'Period of Insurance' is given in the policy schedule and 'Territorial Limits' is defined in the 'Definitions' section of the policy.

The policy then sets out the perils covered which are taken from the clause 21.2.1. These have already been considered in detail in Chapter 9.

The provisos

Proviso 1 together with the policy schedule indicate that the limit of indemnity applies in respect of any one occurrence or series of occurrences arising out of one cause.

Proviso 2 requires notification to the insurer 'within 21 days of entering into or commencing work under such contract or agreement whichever is the sooner together with full details of the contract'. This is necessary as in the case of an annual policy the insurer must have details of every contract where this cover is required, as a survey may be required before cover can be considered which may result in the operation of the next proviso.

Proviso 3 clearly gives the insurer the right to come off cover or impose underwriting terms.

Proviso 4 clearly gives the insured contractor or the employer the right to refuse to accept the underwriting terms just mentioned and failure to accept such terms will be taken as a refusal, the time limit being 14 days from the date of issue of the quotation. While this cover is for the protection of the employer, the contractor can be involved if, for example, the insurer requires the work to be carried out in a certain way such as the shoring of third party property or the use of a certain type of shoring.

Employer

This Section of the policy has a definition of 'Employer'. It reads:

> For the purpose of this Section Employer shall mean any person firm company ministry or authority named as the Employer in the contract or agreement entered into by the Insured.

Presumably the purpose of this definition is to indicate that any person or organisation who commissions work of construction can be the employer in this insurance contract.

Exceptions

Exception 1 – the contract exceptions

Exception 1 follows the exceptions listed in clause 21.2.1 but the wording is slightly different and this calls for comment.

Paragraph (a) reads:

> caused by the negligence omission or default of the Insured or any agent or Employee of the Insured or of any sub-contractor or his employees or agents

Clause 21.2.1 merely refers to the contractor's liability under clause 20.2, and since the amendment in 1986 this clause uses the phrase 'negligence, breach of statutory duty, omission or default of the Contractor [etc.]' so the policy strictly speaking does not cover the requirements of clause 21.2.1 as breach of statutory duty should be excluded as well as negligence, omission or default.

Paragraph (b) reads:

which is attributable to errors or omissions in the planning or the designing of the Contract Works

While the word 'planning' is additional to the wording of the contract clause, this would not make any difference to the operation of the exclusion.

Paragraph (c) reads:

arising from Damage which could reasonably be foreseen to be inevitable having regard to the nature of the work to be executed or the manner of its execution

This exception follows the wording of the clause.

Paragraph (d) reads:

arising from Damage to property which is at the risk of the Employer under the terms of the contract or agreement

While the wording is different from that in the clause the interpretation must be the same.

Paragraph (e) reads:

arising from Contractual Liability

Now 'contractual liability' is defined in the 'Definitions' of the policy as meaning liability which attaches by virtue of a contract or agreement but which would not have attached in the absence of such contract or agreement. This would not apply to the construction contract between the employer and the contractor, but would avoid liability under other contracts which the employer may have entered into which increase his common law liability to the owners or tenants of adjacent property. Such risks might be insurable, but the insurer wants the opportunity to assess this risk if he is asked to insure it. Most insurers will also exclude the liability risk arising from ownership, possession or use by or on behalf of the insured of any mechanically propelled vehicle in circumstances necessitating compulsory third party insurance under statute.

Paragraph (f) reads:

arising from Damage occasioned by pressure waves caused by aircraft or other aerial devices travelling at sonic or supersonic speeds.

This exception is self explanatory and in accordance with the fifty exception in clause 21.2.1. War risks and nuclear risks are excluded under the general exceptions of the policy.

Exception 2 – early shoring required

Exception 2 reads:

if the contract or agreement specifies that shoring of any building or structure is required and such shoring is necessary within 35 days of commencement of the contract or agreement.

Presumably the insurer requires time to investigate the risk hence this exception.

Exception 3 – the hazardous risks

Exception 3 reads:

> against any expense liability loss claim or proceedings arising from
> (a) demolition or partial demolition of any building or structure
> (b) the use of explosives
> (c) tunnelling or piling work
> (d) underpinning
> (e) deliberate dewatering of the site.

This exception clearly lists underwriting terms enabling the insurer to avoid or at least consider these hazardous risks before accepting them, and reinforces the common law duty of disclosure of material facts (the principle of the utmost good faith), see Chapter 2.

Exception 4 – contractual damages

Exception 4 reads:

> in respect of any sum payable under any penalty clause or by reason of breach of contract.

This exception is self explanatory.

Exception – the excess

Exception 5 reads:

> against the Excess specified in the Schedule.

This exception is also self explanatory. Incidentally the excess can be very large.

General exceptions and conditions

The terms set out in Appendix 2 under these headings apply, where stated, to this Section of the policy and have been considered in Chapter 6.

ABI Model Exclusions for 21.2.1 Insurance

The Association of British Insurers has submitted a suggested draft 21.2.1 policy exclusions wording and this is set out in Appendix 10. Whether this wording will be approved remains to be seen.

Chapter 11

Insurance of the works against all risks: clauses 22, 22A and 22B

11.1 Introduction

Article 1 of the 1980 Edition, all versions, provides that 'the Contractor will upon and subject to the Contract Documents carry out and complete the Works shown upon, described by or referred to in those Documents'. Clause 2.1 repeats this requirement and identifies the contract documents mentioned in Article 1. Unless the contract documents provide otherwise, e.g. the 1980 Edition clauses 22B or 22C (see later), the contractor in spite of loss or damage to the works for whatever reason must carry out and complete the works and pay for the expense involved. This follows the basic rule set out in *Charon (Finchley) Ltd* v. *Singer Sewing Machine Co Ltd* (1968), which did not concern the JCT contract but where the court quoted Hudson's *Building and Engineering Contracts*, 9th edition, page 223 as follows:

'Indeed, by virtue of the express undertaking to complete (and in many contracts to maintain for a fixed period after completion) the contractor would be liable to carry out his work again free of charge in the event of some accidental damage occurring before completion even in the absence of any express provision for protection of the risk.'

Therefore the prudent contractor would wish to arrange the widest possible insurance cover for the 'Works'.

Except in the 1980 clause 22B (local authorities versions), the JCT form has for years provided that insurance cover be taken out to deal with loss or damage to the works from fire and special perils, as the insurance world calls this cover: i.e. fire, lightning, explosion, riot and civil commotion, earthquake, aircraft or other aerial devices or articles dropped therefrom, storm, tempest and flood, and bursting or overflowing of water tanks, apparatus or pipes. These perils are now defined in the 1986 Amendment in clause 1.3 as specified perils. This amendment also introduced and defined a requirement for all risks insurance, as set out below.

There are alternatives for the erection of a new building but no choice for the alteration of or extension to an existing building under clause 22, as follows (see footnote [m] of the contract form):

(A) the erection of a new building where the contractor must insure the works against loss or damage by all risks;

(B) the erection of a new building where the employer must insure the works against loss or damage by all risks;

(C) the alteration of or extension to an existing building where the employer must insure:

    (a) the works against all risks

    (b) existing structures and contents owned by him, or for which he is responsible against specified perils.

(The capital letters used above are used in the printed form to denote the optional clause concerned.)

11.2 Clause 22. Insurance of the works

22.1 Clause 22A or clause 22B or clause 22C shall apply whichever clause is stated to apply in the Appendix.

22.2 In clause 22A, 22B, 22C and, so far as relevant, in other clauses of the Conditions the following phrase shall have the meaning given below:

    ...

All Risks Insurance: [n]

insurance which provides cover against any physical loss or damage to work executed and Site Materials but excluding the cost necessary to repair, replace or rectify

1 property which is defective due to

   .1 wear and tear

   .2 obsolescence,

   .3 deterioration, rust or mildew;

[m.1] 2 any work executed or any Site Materials lost or damaged as a result of its own defect in design, plan, specification, material or workmanship or any other work executed which is lost or damaged in consequence thereof where such work relied for its support or stability on such work which was defective;

3 loss or damage caused by or arising from

   .1 any consequence of war, invasion, act of foreign enemy, hostilities (whether war be declared or not), civil war, rebellion, revolution, insurrection, military or usurped power, confiscation, commandeering, nationalisation or requisition or loss or destruction of or damage to any property by or under the order of any government *de jure* or *de facto* or public, municipal or local authority;

   .2 disappearance or shortage if such disappearance or shortage is only revealed when an inventory is made or is not traceable to an identifiable event;

   .3 an Excepted Risk (as defined in clause 1.3);

   and if the Contract is carried out in Northern Ireland

   .4 civil commotion;

   .5 any unlawful, wanton or malicious act committed maliciously by a person or persons acting on behalf of or in connection with an unlawful association; 'unlawful association' shall mean any organisation which is engaged in terrorism and includes an organisation which at any relevant time is a proscribed organisation within the meaning of the Northern Ireland (Emergency Provisions) Act 1973; 'terrorism' means the use of violence for political ends and includes any use of violence for the purpose of putting the public or any section of the public in fear.

The phrase 'Site Materials' is defined in clause 1.3 as follows:

all unfixed materials and goods delivered to, placed on or adjacent to the Works and intended for incorporation therein.

Clauses 22.1 to 22.3 have been added to the 1980 edition by the 1986 Amendment. Clause 22.3 is set out later.

Clause 22.1 is self-explanatory and by a footnote [m] the circumstances under which clauses 22A, 22B and 22C are to be used are the same as in JCT 80 (see above).

Clause 22.2 defines the meaning of all risks insurance. The words 'physical loss' in the insuring clause make it clear that consequential loss is not covered (see Chapter 7). While the footnote [m.1] seems to disallow limitations on the design exclusion (see an explanation of this in Chapter 7), the intention seems to be to disallow any additional exclusions as clauses 22A.1, 22B.1 and 22C.2 (all of which concern all risks insurance), when referring to the cover to be provided, use the phrase 'no less than that defined in clause 22.2 [n] [o.1]', or in the case of clause 22C.2 'no less than that defined in clause 22.2 [n] [o.2]'. However this cannot disallow exclusions setting out risks which are covered by other policies, e.g. motor, marine and engineering risks.

Footnote [n] makes the point that all risks policies vary in their wording and footnote [o.1] states that it may not be possible for insurance to be taken out against certain of the risks covered by the definition of 'all risks insurance'. It says this matter should be arranged between the parties prior to entering into the contract, and either the definition of 'all risks insurance' given in clause 22.2 should be amended or the risks actually covered should replace this definition. In the latter case, clause 22A.1, 22A.3 or 22B.1 (whichever is applicable) and other relevant clauses in which the definition 'all risks insurance' is used should be amended to include the words used to replace this definition, e.g. riot cover may be unobtainable in Northern Ireland. Footnote [o.2] which also applies to clause 22C.2, refers to both all risks and specified perils insurance and makes the same point that footnote [o.1] makes, except that it specifically mentions clause 22C.1 and/or clause 22C.2. Finally footnote [o.2A], which applies to clause 22C.1 states that some employers, e.g. tenants may not be able to fulfil the obligations in clause 22C.1. If so clause 22C.1 should be amended accordingly.

The extent of this insurance now required is not unreasonable but it means that there will be many insured who will have annual blanket or floater policies which do not conform with the revised contract conditions. This is particularly so regarding the extension of design, etc. cover now required. While it should be simple to extend the cover, whether insurers will be prepared to provide the JCT cover for all contracts, whether subject to JCT conditions or not, does not seem to have arisen. Possibly this is due to the fact that according to the *RICS Contracts in Use Survey 1993* JCT Standard Forms continue to dominate the market. Whether taken by number or by contract value, JCT Forms of one sort or another account for 80% of the sample. Similarly whether insureds will refuse to pay on non-JCT contracts for such cover where they do not consider it necessary does not seem to have arisen.

Contractors' all risks policies issued by insurers are not standardised and there is some variation in the way these policies, and particularly the exclusions, are expressed.

The all risks insurance exclusions listed in sub-clause 22.2 have been explained in Chapter 7, bearing in mind the Excepted Risks in clause 1.3 include nuclear risks and sonic waves.

11.3 Sub-contractors

22.3.1 The Contractor where clause 22A applies, and the Employer where either clause 22B or clause 22C applies, shall ensure that the Joint Names Policy referred to in clause 22A.1 or clause 22A.3 or the Joint Names Policies referred to in clause 22B.1 or in clauses 22C.1 and 22C.2 shall

either provide for recognition of each sub-contractor nominated by the Architect as an insured under the relevant Joint Names Policy

or include a waiver by the relevant insurers of any right of subrogation which they may have against any such Nominated Sub-Contractor

in respect of loss or damage by the Specified Perils to the Works and Site Materials where clause 22A or clause 22B or clause 22C.2 applies and, where clause 22C.1 applies, in respect of loss or damage by the Specified Perils to the existing structures (which shall include from the relevant date any relevant part to which clause 18.1.3 refers) together with the contents thereof owned by the Employer or for which he is responsible; and that this recognition or waiver shall continue up to and including the date of issue of the certificate of practical completion of the sub-contract works (as referred to in clause 2.11 of Conditions NSC/C (*Conditions of Nominated Sub-Contract*) or the date of determination of the employment of the Contractor (whether or not the validity of that determination is contested) under clause 27 or clause 28 or clause 28A or, where clause 22C applies, under clause 27 or clause 28 or clause 28A or clause 22C.4.3, whichever is the earlier. The provisions of clause 22.3.1 shall apply also in respect of any Joint Names Policy taken out by the Employer under clause 22A.2 or by the Contractor under clause 22B.2 or under clause 22C.3 in respect of a default by the Employer under clause 22C.2.

22.3.2 Except in respect of the Joint Names Policy referred to in clause 22C.1 (or the Joint Names Policy referred to in clause 22C.3 in respect of a default by the Employer under clause 22C.1) the provisions of clause 22.3.1 in regard to recognition or waiver shall apply to Domestic Sub-Contractors. Such recognition or waiver for Domestic Sub-Contractors shall continue up to and including the date of issue of any certificate or other document which states that the domestic sub-contract works are practically complete or the date of determination of the employment of the Contractor as referred to in clause 22.3.1, whichever is the earlier.

This clause 22.3 should be read carefully and particular notice taken of the following points as frequently sub-contractors damage the works and the question of recovery arises.

(1) The contractor, where clause 22A applies, and the employer, where either 22B or 22C applies, have to ensure that the joint names policies mentioned in those clauses either provide for recognition of each sub-contractor nominated by the architect as an insured under those policies, *or* include a waiver by the relevant insurers of any right of subrogation which they may have against any such nominated sub-contractor *but only in respect of loss or damage by the specified perils* to the works and site materials or the existing structures as the case may be. Specified perils are the same as the old 'clause 22 perils' (see the fire and special perils listed earlier).

(2) This means that even when all risks insurance is required by contract, i.e. in clause 22A or 22B or 22C.2, only the specified perils cover of the all risks insurance is to operate so far as nominated sub-contractors are concerned. Apparently it was not considered possible to get insurers generally to extend the all risks insurance of the works to provide the all risks cover to nominated sub-contractors. However, with the agreement of insurers the 1986 version of NSC/4 and 4a now NSC/C (the Nominated Sub-contract, see later) provides for the nominated sub-contractors to obtain the benefit of the major part of the cover given to the employer and contractor jointly under the main contract 1986 joint names policy. This is the explanation given in the *Guide to the Amendments to the Insurance and Related Liability Provisions* 1986 issued with Practice Note 22.

(3) As regards existing structures to which clause 22C.1 applies, the application of clause 18.1.3 concerning partial possession by the employer means that the part of the works of which the employer is to take possession (the relevant part) from the relevant date is in effect no longer regarded as 'the Works' but as an 'existing structure' and therefore subject to the joint names policy taken out by the employer under clause 22C.1. The contents of the existing structures owned by the employer, or for which he is responsible, are also included in the joint names policy cover.

(4) This recognition of each nominated sub-contractor under the policy or waiver of the right of subrogation is to continue up to and including the date of issue of the certificate of practical completion of the *sub-contract works* (as referred to in clause 2.11 of sub-contract NSC/C) or the date of determination of the employment of the contractor (whether or not the validity of that determination is contested) under the earliest of the clauses listed.

(5) Except in respect of the joint names policy referred to in clause 22C.1 (or that referred to in clause 22C.3 in respect of a default by the employer under clause 22C.1) the provisions of clause 22.3.1 regarding recognition or waiver shall apply to domestic sub-contractors up to and including the date of issue of any certificate or other document which states the domestic sub-contract works are practically complete, or the date of determination of the employment of the contractor as referred to in clause 22.3.1, whichever is the earlier.

To summarise the position under clause 22.3, the two sub-clauses 22.3.1 (concerning nominated sub-contractors) and 22.3.2 (concerning domestic sub-contractors) protect sub-contractors against specified perils either as an insured person or by waiver of subrogation rights by insurers, but domestic sub-contractors do not get this benefit under sub-clause 22C.1 for existing structures and contents. The cases of *Welsh Health Technical Services Organisation* v. *Haden Young* (1986) and *Norwich City Council* v. *Harvey and Briggs Amasco* (1989) on JCT terms operating before the 1986 Amendment indicate that the duty of care otherwise owed by the sub-contractor defendant to the plaintiff employer can be qualified by the terms of the contract between the parties (although the employer never intended privity of contract between himself and the sub-contractor imposing liability for fire and other special perils). By the terms concerned in those cases the plaintiff employer accepted the risk of fire and other special perils to its property under clauses 20B and 20C (now 22B and 22C).

However, the position under the 1986 Amendment is clearer, but it leaves all sub-contractors responsible for losses, other than by specified perils against which there is no contractual requirement to insure, and domestic sub-contractors carry the risk of all loss or damage to existing structures and their contents. Incidentally, insurers refuse to give sub-contractors all risks cover under the main contractor's or employer's CAR policy, particularly under annual open cover policies, as the materials they use are often attractive to thieves and the care taken in storage on site can vary. Furthermore, if the theft and malicious damage risks will not affect the sub-contractor's record it may influence his security attitude. Also, the sub-contractor's claims record will directly affect the main contractor's insurance record if they were included within his insurance cover.

11.4 Clause 22A. Erection of new buildings – all risks insurance of the works by the contractor

Joint names policy

22A.1 The Contractor shall take out and maintain a Joint Names Policy for All Risks Insurance for cover no less than that defined in clause 22.2 [n] [o.1] for the full reinstatement value of the Works (plus the percentage, if any, to cover professional fees stated in the Appendix) and shall (subject to clause 18.1.3) maintain such Joint Names Policy up to and including the date of issue of the certificate of Practical Completion or up to and including the date of determination of the employment of the Contractor under clause 27 or clause 28 or clause 28A (whether or not the validity of that determination is contested), whichever is the earlier.

Clause 22A.1 requires the contractor to take out and maintain a joint names policy for all risks insurance for work executed and site materials as it states 'no less than that defined in clause 22.2 above' and it should be noted that clause 1.3 includes a definition of site materials. The cover has to be for the full reinstatement value of the works (plus the percentage, if any, to cover professional fees). If the insurance money is not adequate to cover the 'full reinstatement value' the shortfall would be the responsibility of the contractor as he has a liability to complete the works (see the beginning of this chapter).

Consideration should be given to obtaining a policy which automatically provides for an uplift in the sum insured by a percentage which will adequately reflect the increase in costs arising from inflation in respect of building costs and professional fees. The contract price is rarely, if ever, sufficient, so a type of escalation clause just described is important.

The liability to insure continues up to the date of issue of the certificate of practical completion (including that date) or up to and including the date of determination of the employment of the contractor under clauses 27, 28 or clause 28A whichever is the earlier. However, this is subject to clause 18.1 which indicates that the sum insured is reduced by the full value of any part possession taken over by the employer as from the date on which the employer takes possession.

> 22A.2 The Joint Names Policy referred to in clause 22A.1 shall be taken out with insurers approved by the Employer and the Contractor shall send to the Architect for deposit with the Employer that Policy and the premium receipt therefor and also any relevant endorsement or endorsements thereof as may be required to comply with the obligation to maintain that Policy set out in clause 22A.1 and the premium receipts therefor. If the Contractor defaults in taking out or in maintaining the Joint Names Policy as required by clauses 22A.1 and 22A.2 the Employer may himself take out and maintain a Joint Names Policy against any risk in respect of which the default shall have occurred and a sum or sums equivalent to the amount paid or payable by him in respect of premiums therefor may be deducted by him from any monies due or to become due to the Contractor under this Contract or such amount may be recoverable by the Employer from the Contractor as a debt.

It is important to note that a side heading to this sub-clause reads 'Single policy – insurers approved by Employer – failure by Contractor to insure'. Consequently, in the event of the completion date being extended, then it is essential to extend this single policy accordingly.

This is the usual sub-clause which appears in most construction contracts requiring insurers to be approved by the employer and the policy, premium receipt and relevant endorsement(s) to be sent to the employer via the architect. Thus any statutory authorised insurer could be approved.

The employer's authority to insure if the contractor fails to do so is self-explanatory. There is a specimen all risks policy in the JCT contract, as far as the insuring clause and exclusions are concerned so there is clearly a test to be applied, whereas in the past the matter was usually left to an insurance broker specialising in construction work to arrange appropriate cover.

Annual policies

22A.3.1 If the Contractor independently of his obligations under this Contract maintains a policy of insurance which provides (*inter alia*) All Risks Insurance for cover no less than that defined in clause 22.2 for the full reinstatement value of the Works (plus the percentage, if any, to cover professional fees stated in the Appendix) then the maintenance by the Contractor of such policy shall, if the policy is a Joint Names Policy in respect of the aforesaid Works, be a discharge of the Contractor's obligation to take out and maintain a Joint Names Policy under clause 22A.1. If and so long as the Contractor is able to send to the Architect for inspection by the Employer as and when he is reasonably required to do so by the Employer documentary evidence that such a policy is being maintained then the Contractor shall be discharged from his obligation under clause 22A.2 to deposit the policy and the premium recept with the Employer but on any occasion the Employer may (but not unreasonably or vexatiously) require to have sent to the Architect for inspection by the Employer the policy to which clause 22A.3.1 refers and the premium receipts therefor. The annual renewal date, as supplied by the Contractor, of the insurance referred to in clause 22A.3.1 is stated in the Appendix.

The side heading to this sub-clause reads 'Use of annual policy maintained by Contractor – alternative to use of clause 22A.2'.

Where the contractor's annual policy is used, the last sentence of this clause provides for the contractor to supply the employer with the annual renewal date which is inserted in the appendix so that the employer is aware of the date the policy has to be renewed. The purpose of this sub-clause is to use the contractor's annual all risks policy endorsed in the joint names of the employer as well as the contractor in respect of the works concerned and so discharge the contractor's obligation under clause 22A.1. By sending documentary evidence of the maintenance of the policy to the architect for inspection by the employer the contractor complies with his obligation under clause 22A.2.

Default in insuring

22A.3.2 The provisions of 22A.2 shall apply in regard to any default in taking out or in maintaining insurance under clause 22A.3.1.

Under this sub-section the same provisions apply in respect of the employer's rights to insure, should the contractor default in respect of his existing annual insurance mentioned in sub-clause 22A.3.1.

Making claims

22A.4.1 If any loss or damage affecting work executed or any part thereof or any Site Materials is occasioned by any one or more of the risks covered by the Joint Names Policy referred to in clause 22A.1 or clause 22A.2 or clause 22A.3 then, upon discovering the said loss or damage, the Contractor shall forthwith give notice in writing both to the Architect and to the Employer of the extent, nature and location thereof.

22A.4.2 The occurrence of such loss or damage shall be disregarded in computing any amounts payable to the Contractor under or by virtue of this Contract.

22A.4.3 After inspection required by the insurers in respect of a claim under the Joint Names Policy referred to in clause 22A.1 or clause 22A.2 or clause 22A.3 has been completed the Contractor with due diligence shall restore such work damaged, replace or repair any such Site Materials which have been lost or damaged, remove and dispose of any debris and proceed with the carrying out and completion of the Works.

22A.4.4 The Contractor, for himself and for all Nominated and Domestic Sub-Contractors who are, pursuant to clause 22.3, recognised as an insured under the Joint Names policy referred to in clause 22A.1 or clause 22A.2 or clause 22A.3, shall authorise the insurers to pay all monies from such insurance in respect of the loss or damage referred to in clause 22A.4.1 to the Employer. The Employer shall pay all such monies (less only the percentage, if any, to cover professional fees stated in the Appendix) to the Contractor by instalments under certificates of the Architect issued at the Period of Interim Certificates.

22A.4.5 The Contractor shall not be entitled to any payment in respect of the restoration, replacement or repair of such loss or damage and (when required) the removal and disposal of debris other than the monies received under the aforesaid insurance.

These remaining sub-clauses deal with insurance claims for loss or damage to the works, etc.

Sub-clause 22A.4.1 concerns the requirement of the contractor to notify and give details to the architect and the employer if such loss or damage occurs which is covered by the joint names policy. Otherwise the architect and employer may know nothing about it.

Sub-clause 22A.4.2 provides that the occurrence of loss or damage is to be disregarded in computing amounts payable to the contractor. Thus the contractor must be paid for all such work as though no loss or damage had occurred. This means the contractor is entitled to payment for the reinstatement work from the insurance money received under the joint names policy as well as for the work he had originally done in accordance with the contract but which had been subsequently lost or damaged.

Sub-clause 22A.4.3 emphasises the fact that the contractor has to restore damaged work or replace or repair damaged or lost site materials in respect of a claim under the joint names policy, but only after the insurers have carried out any inspection they may require.

Sub-clauses 22A.4.4 and 22A.4.5 indicate that adequate policy cover and sum insured are vital since only insurance monies (less the percentage, if any, to cover professional fees stated in the appendix) shall be available for payments to the employer, and then by certificates (usually monthly) of the architect, to the contractor in respect of restoration of the damage, i.e. by instalments. Note that in sub-clause 22A.4.4 sub-contractors can be considered joint insureds or subrogation rights waived by insurers but only in respect of the specified perils as stated in clause 22.3.

Theory and practice

The wording of clause 22A.4.3 and also for clauses 22B.3.3 and 22C.4.4.1 (see below and Chapter 13) in practice is not adhered to as precisely as they may suggest. For example, insurers may not wish to inspect for minor damage or may even delay for many reasons. In the meantime the contractor will be under pressure to proceed with repairs, particularly under clauses 22B and 22C contracts as such repairs are deemed to be clause 13.2 variations, and therefore give entitlement to clause 25 extension of time and possibly a clause 26 loss and expense claim. In practice in clause 22 the insurers must be given a reasonable opportunity to inspect, but if they do not respond the contractor must proceed with due diligence to restore such work, damage, etc. on the basis that no inspection is required by the insurers. Incidentally in paying the employer under the joint names policy, the insurers' liability would be restricted to the cost of effecting repairs and would not include consequential loss, such as the contractor's delay loss and expense claim. Practice suggests that experience also differs significantly from wording when clauses 22A.4.4 and 22A.4.5 are considered.

The wording suggests that, where clause 22A applies, a contractor can only receive payment from the joint names policy (and that position is supported by practice) but the wording further suggests that the insurers pay the employer and the employer then pays the contractor by instalments under special certificates issued by the architect. Practice suggests that, for all but possibly significant claims, the contractor deals directly with and obtains payment directly from the joint names insurers and neither the employer nor the architect become involved.

11.5 Clause 22B. Erection of new buildings – all risks insurance of the works by the employer

22B.1 The Employer shall take out and maintain a Joint Names Policy for All Risks Insurance for cover no less than that defined in clause 22.2 [n] [o.1] for the full reinstatement value of the Works (plus the percentage, if any, to cover professional fees stated in the Appendix) and shall (subject to clause 18.1.3) maintain such Joint Names Policy up to and including the date of issue of the certificate of Practical Completion or up to and including the date of determination of the employment of the Contractor under clause 27 or clause 28 or clause 28A (whether or not the validity of that determination is contested) whichever is the earlier.

22B.2   The Employer shall, as and when reasonably required to do so by the Contractor, produce documentary evidence and receipts showing that the Joint Names Policy required under clause 22B.1 has been taken out and is being maintained. If the Employer defaults in taking out or in maintaining the Joint Names Policy required under clause 22B.1 then the Contractor may himself take out and maintain a Joint Names Policy against any risk in respect of which a default shall have occurred and a sum or sums equivalent to the amount paid or payable by him in respect of the premiums therefor shall be added to the Contract Sum.

22B.3.1 If any loss or damage affecting work executed or any part thereof or any Site Materials is occasioned by any one or more of the risks covered by the Joint Names Policy referred to in clause 22B.1 or clause 22B.2 then, upon discovering the said loss or damage, the Contractor shall forthwith give notice in writing both to the Architect and to the Employer of the extent, nature and location thereof.

22B.3.2 The occurrence of such loss or damage shall be disregarded in computing any amounts payable to the Contractor under or by virtue of this Contract.

22B.3.3 After any inspection required by the insurers in respect of a claim under the Joint Names Policy referred to in clause 22B.1 or clause 22B.2 has been completed the Contractor with due diligence shall restore such work damaged, replace or repair any such Site Materials which have been lost or damaged, remove and dispose of any debris and proceed with the carrying out and completion of the Works.

22B.3.4 The Contractor, for himself and for all Nominated and Domestic Sub-Contractors who are, pursuant to clause 22.3, recognised as an insured under the Joint Names Policy referred to in clause 22B.1 or clause 22B.2, shall authorise the insurers to pay all monies from such insurance in respect of the loss or damage referred to in clause 22B.3.1 to the Employer.

22B.3.5 The restoration, replacement or repair of such loss or damage and (when required) the removal and disposal of debris shall be treated as if they were a Variation required by an instruction of the Architect under clause 13.2.

Comparison with clause 22A

This clause 22B was drafted in 1986 in similar terms to clause 22A, but of course with the employer not the contractor taking out the joint names policy for all risks insurance.

In sub-clause 22B.1 there is no longer provision for the employer to take the 'sole risk' of loss or damage. Instead there is an obligation on the employer, in both the private and local authorities' editions, to take out the insurance mentioned in the previous paragraph. This was not previously required in the local authorities' edition. The insurance has to conform with the requirements of clause 22.2. Because of the joint names policy the necessity for both parties to insure their interest is now avoided as the rights of subrogation by the insurer against the contractor cannot exist under this type of policy. Prior to the 1986 Amendment the contractor would have to insure the works, etc. against the non-listed perils as the cover required from the employer only applied to fire and specific perils.

The omitted 'sole risk' clause was one of the reasons which led to the employer being held liable for damage caused by negligence of the contractor in the case of *Archdale & Co Ltd* v. *Comservices Ltd* (1954) and *Scottish Special Housing Association* v. *Wimpey* (1986) but more details of these cases will be given in the commentary on clause 22C in Chapter 13.

In the 1986 private version of clause 22B.2 the contractor is entitled to proof from the employer that the joint names policy exists and if the employer defaults in his obligation, the contractor may take out and maintain this policy. In the local authorities' version the contractor has no such right and clause 22B.2 is marked 'Number not used'.

It has been suggested that if the contractor did not use this right to insure he would lose little, as he has been paid to date and the determination rights under clause 28 are generous to the contractor. Also the retention funds under clause 30.5.1 might constitute a trust fund in the event of the employer's insolvency.

Clause 22B.3 contains five sub-clauses dealing with:

- insurance claims for loss or damage to the works, etc. requiring notice and details by the contractor as in sub-clause 22A.4.1;
- payment to the contractor as in sub-clause 22A.4.2;
- restoration of work after insurers' inspection as in sub-clause 22A.4.3;
- payment of insurance monies to the employer as in sub-clause 22A.4.4;
- treatment of repair and (if required) removal of debris as if they were a variation under clause 13.2 in contrast to clause 22A.4.5 which limits payment to the amount of the insurance.

All sub-contractors are included in accordance with clause 22.3.

Employer's risk

No change was made to the principle in the 1980 clause 22B that the employer pays for the restoration, etc. of loss or damage to the works as if it were a variation. Thus any shortfall in insurance payment in meeting the cost of paying for the variation is borne by the employer. In clause 22A the shortfall is borne by the contractor.

It might well be asked why the employer should ever accept this responsibility, and, so far as the private edition is concerned where the employer only uses the contract once or twice in a lifetime, the optional sub-clause 22B is probably not used. However, the regular user of the private edition, e.g. a building developer or a local authority who can arrange insurance cover at more advantageous rates than most contractors, may decide to use this sub-clause.

Duncan Wallace in *Hudson's Building and Engineering Contracts*, 11th edition (1995) has the following to say about this subject:

'The optional insurance of the works clause, whereby the owner undertakes to insure and express provision is made for all risk of damage to the works to pass to him, is such as no adviser of the owner should accept, since, apart from insurance, it overturns and reverses the substantive contractual liability of the contractor.'

Presumably this refers to the contractor's obligation under clause 2.1 to complete the contract, and the assumption that clause 22B gives no inducement to the contractor to exercise reasonable care in protecting the works from damage.

If it is accepted that this optional sub-clause is only used where the employer is large and influential enough to obtain better insurance rates than the contractor because of the bulk business offered, then it would follow that in practice the contractor would be very anxious not to offend such a large and influential customer from whom he would expect to obtain further work or at least the opportunity to tender for further contracts. This must lessen the effect of the adverse features mentioned in the previous paragraph.

Theory and practice

As in the case of the wording of Clause 22A, it appears that practice can often vary from wording.

Where clause 22B applies, it is the intention of the wording that the repair be treated as if it were a variation required under clause 13.2. The effect of this should be that the repairs are valued by the quantity surveyor in accordance with the normal contract arrangements and the contractor should be paid through the contract, leaving the employer to recover from the joint names insurer. This is compatible with the position under clause 22B.3.4 whereby the contractor, for himself and for all nominated and domestic sub-contractors, is required to authorise the joint names insurers to pay all claims to the employer.

In practice, however, it is usually the case that, once again, the contractor is left to deal directly with the joint names insurer, or at least loss adjusters acting for them. Although payment is not usually made direct from joint names insurers to contractors, nonetheless, in practice, it is the insurers or loss adjusters who negotiate agreement on quantum instead of, as should be the case, the quantity surveyor valuing the works in accordance with the contract conditions. Where clause 22B applies, the contractor's entitlement to payment arises under the contract and is not conditional upon the acceptance of a claim by the joint names insurers, that risk rests with the employer.

The model clauses

Some local authority employers prefer to be responsible for insurance themselves and when the JCT issued its amendments to the insurance clauses in 1986, it published *Practice Note 22 and Guide to the Amendments* which included several model clauses which may be used instead of clause 22B (see Appendix 6).

These model clauses allow the employer to take either 'the risk' or 'the sole risk' of loss or damage to the works, but do not require insurance. This difference in risks means that, in the case of the contractor's or sub-contractor's negligence causing such loss, etc., the employer will still have to pay where 'sole risk' applies but, where merely 'risk' applies, he will not have to pay.

Where the model clauses are used, the repair must still be treated as if it were a variation required under clause 13.2.

Chapter 12

Fire and special (or extended) perils policy

12.1 Introduction

It will be appreciated that clauses 22.3 and 22C.1 of the current JCT Form (1980 Edition with the 1986 Amendment), by referring to specified perils and defining them in clause 1.3, acknowledge the acceptance of a more limited type of cover than the contractors' all risks policy. The sub-contractor in clause 22.3 will no doubt take advantage of that part of the main contractor's CAR policy which equates with the specified perils as defined. Similarly where the employer is providing the all risks cover for the works the sub-contractor will take advantage of the employer's policy. However, where clause 22C.1 applies the employer may already have a fire and special perils policy covering the existing structures and their contents and special arrangements will have to be made for the specified perils cover required for the nominated sub-contractor. No such cover is required for the domestic sub-contractor. If the employer has no insurance or only a fire policy covering the existing structures and their contents then he will have to arrange a policy in the form described in this chapter, to cover both himself and any nominated sub-contractor. It has to be remembered that the CAR policy does not cover existing structures and their contents, it only covers the works etc.

The fire and special perils (or additional or extended perils) policy is a material damage insurance. The standard fire policy only covers fire, lightning and (limited) explosion risks, but the JCT contract requires cover for the following additional perils under clause 22 where specified perils are called for:

- unlimited explosion from any insurable cause;
- riot and civil commotion;
- earthquake;
- aircraft or other aerial devices or articles dropped therefrom;
- storm, tempest and flood;
- bursting or overflowing of water tanks, apparatus and pipes.

A fire policy carrying the wording extending the cover (see Appendix 1) includes the perils listed in clause 1.3 of the JCT contract.

Until 1985 the Fire Offices Committee controlled the tariff insurers and required a Standard Fire Policy, the wording of which was followed by the non-tariff insurers. This organisation ceased to exist in 1985 and was succeeded by the Property Committee of the Association of British Insurers (ABI) who issued a recommended wording to its members under the title 'Standard Fire Policy (Material Damage')'. The basic features of the old Standard Fire Policy are retained by the ABI wording. However, the old form of fire policy will continue in existence because these policies are renewed each year.

At this stage it is necessary to point out that a fire and special perils policy of the type mentioned would have exclusions, some of which are common to all the perils and the remainder applicable to only certain of the perils. The initial letters CAR will be used for the contractors' all risks policy where mentioned. There are two general points to be made:

- The exclusions common to all the perils concern war and kindred risks, nuclear risks and consequential loss (see Appendix 1, special conditions 4(a)(ii), 4(b) and 4(d)). Consequential losses are the subject matter of separate policies and were considered when dealing with the CAR policy. The other special conditions are considered separately or with the peril to which they apply.
- Any conditions of average in the policy would apply but the penalties which would arise from the operation of such conditions can be avoided by insuring for an adequate sum.

12.2 Detailed consideration

Consideration will now be given to the various perils and the relative conditions and exclusions. These perils will be discussed in the order set out in Appendix 1 (except that unlimited explosion protection will be dealt with after discussing limited explosion cover).

Standard Fire Policy perils

In the first place, the basic cover of the Standard Fire Policy (not given in Appendix 1), i.e. fire, lightning and limited explosion, will be dealt with.

(1) Subject to some general exclusions (see below), the fire cover is in respect of fire damage **not** caused by:

- explosion resulting from fire;
- earthquake or subterranean fire;
- spontaneous fermentation or heating of the property insured; or
- property undergoing any heating process or any process involving the application of heat.

It should be noted that in respect of the 'fermentation' and 'heating' exclusions only the property directly involved is excluded, any damage arising from the spread of such a fire is covered.

(2) The peril of lightning needs no comment.

(3) There is limited explosion cover in respect of domestic boilers and in respect of gas used for domestic purposes. Explosion caused by earthquake or subterranean fire is excluded. Clearly the limited explosion cover is of little value to the commercial insured save that an explosion of gas used for cooking or in heating a mess hut would arguably be gas used for domestic purposes.

Standard Fire Policy exclusions and conditions

There are general exclusions and conditions in the Standard Fire Policy.

Exclusions

The peril exclusions are:

- General exclusion 1 is riot, civil commotion, war, invasion, act of foreign enemy, hostilities (whether war be declared or not) civil war, rebellion, revolution, insurrection or military or usurped power.
- Nuclear risks are also excluded at 2 (a) and (b).
- Exclusion 3 is in respect of 'Terrorism'. This is in respect of Northern Ireland in the standard document but as a matter of routine a similar exclusion is now added in relation to the UK mainland. The full wording is as set out in Chapter 7 (exception 9 of the CAR policy). It should be noted that the latter clause of this exclusion reverses the onus of proof – the insured has to counter the assertion of the insurer.
- There is an important exclusion 4 regarding damage caused by pollution or contamination.
- General exclusion 5 excludes damage covered by a marine policy.
- Exclusion 6 excludes property more specifically insured by or on behalf of the insured.
- Consequential losses are excluded at 7.

Conditions

The General Conditions of the Standard Fire Policy deal with:

(1) Misdescription which is a warning of the principle of utmost good faith.
(2) Alteration which deals with the removal, increase in risk and change of interest.
(3) Warranties (see Chapter 2) which points out the consequences of a breach of warranty.
(4) The insured is required to take reasonable precautions to prevent damage.

A final section of the policy is the conditions concerned with the making of a claim:

(1) Sets out the action required of the insured.
(2) Refers to fraudulent claims.
(3) Deals with the insurers' right to reinstate the damaged property.
(4) Sets out insurers' rights after a claim.
(5) Also refers to contribution and average (see Chapter 6 and Chapter 7 under the heading 'Additional covers' – 5, Tools, plant, equipment and temporary buildings).
(6) Is concerned with the right of subrogation (see Chapter 6).
(7) Deals with arbitration which, following an undertaking given by insurers through a Law Reform Committee Report (Cmnd 62) in 1957, only applies to the amount to be paid under the policy, not to a dispute on liability.

The limitations on the fire cover given by the CAR policy are less than those in the fire a special perils policy in that there are no limitations which apply only to the fire cover in the CAR policy as there are in the fire and special perils policy.

The procedure for extending the Standard Fire Policy to include special perils

The procedure for extending the fire policy to include special perils is to establish a basic 'extension' wording as in Appendix 1 and then to insert at the appropriate point the reference to the perils which are required using the words set out in the Appendix. In Appendix 1 only the perils required by the JCT contract have been included.

Explosion

To obtain the unlimited explosion cover required by clause 22, two insurances are required:

- the extension to the fire policy as Part C1 of appendix 1 which gives very wide cover but excludes 'explosion of steam pressure vessels' and the like for which
- an engineering insurance is required.

This exclusion does not appear in respect of explosion cover given by the CAR policy. Thus the only additional cover required is for the steam pressure vessel itself. Note that sonic waves are excluded.

Aircraft, etc.

Cover against destruction or damage caused by aircraft and other aerial devices or articles dropped therefrom excludes destruction or damage occasioned by pressure waves caused by aircraft and other aerial devices

Storm and flood

Storm tempest and flood cover is referred to in the JCT contract but as
'tempest' only means a severe storm it is now omitted from the extension
wording in Appendix 1.

Storm and flood cover usually excludes damage caused:

- by frost or by subsidence or by landslip,
- to fences, gates and hoardings,
- to moveable property in the open, or in transit, and
- solely by a change in the water table level.

An excess is usually applicable in the case of storm and flood to avoid the
small claims which are administratively expensive to handle and to give the
insured a financial interest in protecting the property from these perils.
According to *Oddy* v. *Phoenix Assurance Co Ltd* (1996), 'storm' means some
sort of violent wind usually accompanied by rain, hail or snow. It does not
mean persistent bad weather or heavy rain, nor persistent rain by itself.
However, in the Scottish case of *Glasgow Training Group (Motor Trade) Ltd* v.
Lombard Continental plc (1988) the collapse of a building roof caused by a
heavy fall of snow was held to be damage caused by a 'storm' covered by an
insurance policy. The word 'storm' was considered to be properly used
where the precipitation was of an extreme or unusual intensity, particularly
as the policy had to be construed *contra proferentem.*

An excess of water, snow or hail descending from the sky with a violent
wind is treated as 'storm and tempest', whereas the loss caused to property,
etc., by an overflowing river or a high tide washing over a promenade, is
treated as a 'flood' loss. The insurers in Appendix 1 have indirectly stated
what they consider to be 'flood' by the basic 'storm' extension wording (not
shown) as distinct from the 'storm and flood' extension under discussion. In
the 'storm' extension wording they exclude damage caused by

'(i) the escape of water from the normal confines of any natural or artificial
      water course lake reservoir canal or dam
(ii) inundation from the sea'

By implication this is their definition of 'flood'.

It is worth noting in *Young* v. *Sun Alliance and London Insurance Ltd* (1976) in
a case where water seeped up to cover the floor of a lavatory of a private
dwelling, the judge ruled that this was not 'flood' in the terms of the policy.
An element of 'suddenness' was required.

Insurers have decided that they will not consider 'frost' as storm and
exclude it from the cover.

On the meaning of 'subsidence', Lewis J in *David Allen & Sons (Billposting)
Ltd* v. *Drysdale* (1939) said:

'"Subsidence" means sinking, that is to say, movement in a vertical
direction as opposed to "settlement", which means movement in a lateral

direction, but I am of the opinion that the word "subsidence" in this policy covers both subsidence in the sense in which I have defined it, and also settlement.'

Consequently, subsidence incorporates settlement. Most insurers therefore regard ground heave, i.e. movement upwards, in a similar way to subsidence. 'Landslip' was defined by Veale J in *Oddy* v. *Phoenix Assurance Co Ltd* (1966) as follows:

'Landslip is a small landslide. One can perhaps define a landslide in different ways but the accepted definition was a rapid downward movement (under the influence of gravity) of a mass of rock or earth on a slope.'

Most insurers will include an excess in respect of these perils of storm and flood for each and every loss. Unless the employer, engineer or architect is prepared to accept the exclusions, the only alternative is to effect a CAR policy which does not contain these exclusions, although it does contain a larger excess.

Escape of water

The risk of bursting or overflowing of water tanks, apparatus and pipes is now given under the wide peril 'Escape of Water' which excludes:

- damage by water discharged or leaking from any automatic sprinkler installation
- damage in respect of any building which is empty or not in use

These exclusions are justified for the following reasons:

- the insured should remedy the defect as soon as he is aware of it;
- there is a sprinkler leakage policy available which provides the necessary cover, and in the case of a fault in installation the responsibility lies with the sprinkler engineering firm installing the apparatus.

There is an excess, the main purpose of this being to exclude small claims which, in some cases, could be regarded as normal wear and tear. However, once again, a CAR policy would avoid these two exclusions but would have a larger excess. (Note that loss or damage caused by water used for the extinction of fire would form part of a fire loss.)

In *Computer & Systems Engineering plc* v. *John Lelliott (Ilford) Ltd and Others* (1989) it was decided that the contractor, under clause 22C of JCT 80, bears the risk of damage from a sub-contractor's negligent fracture of a water pipe. In contribution proceedings within the main action the sub-contractor was ordered to indemnify the contractor.

Computer & Systems Engineering plc entered into a JCT 80 contract (pre-1986) with the contractor John Lelliott (Ilford) Ltd for work at the former's premises. EWG Stoddart Ltd was the sub-contractor required to install metal purlins. While a purlin was being lifted into position by the sub-contractor, it fell and fractured one of the pipes of the sprinkler system. Water emerged at high pressure and damaged the goods of the employer. The incident was caused by the negligence of the sub-contractor which was joined in the action as second defendant and by the contractor, the first defendant, as third party.

It was claimed that the employer could not recover damages against the contractor or the sub-contractor by reason of clause 22C.1 of JCT 80. This reads:

> The existing structures together with the contents thereof ... shall be at the sole risk of the employer as regards loss or damage by the clause 22 perils.

The clause 22 perils are defined in clause 1.3 as including:

> flood, bursting or overflowing of water tanks, apparatus or pipes ...

It was contended by the contractor and subcontractor that, under the contract, the incident was a 'bursting or overflowing of water tanks, apparatus or pipes' and therefore the responsibility of the employer.

It was held that:

- The employers were not liable under clause 22.
- 'The words are to be given their ordinary meaning, bearing in mind the context in which they occur and the general nature of the contractual provision.'
- 'The court should not be astute to construe a contractual term which in effect completely absolves a wholly and solely guilty tortfeasor from the consequences of his wrong.'
- 'Bursting or overflowing' were to be construed intransitively – involving some disruption of a pipe from within: *Commonwealth Smelting Ltd* v. *Guardian Royal Exchange* (1986).
- John Lelliot (Ilford) Ltd was liable in damages to Computer & Systems Engineering plc but was entitled to an indemnity from the subcontractors, EWG Stoddard Ltd.

12.3 Floating policies for buildings in course of erection

If the contractor is generally engaged in one type of work, e.g. housebuilding or roadworks, it is possible to arrange a floating policy. The insurance can be in respect of fire, or fire and special perils, and covers the contract works, temporary works, tools and employees' effects. An annual declaration is submitted of the value of the work undertaken, which is then rated according to its category, thus taking care of work undertaken outside the normal type. This policy is not subject to average, but, if the sum insured is inadequate, it could affect claims because the money available to effect repairs or reinstatement would be inadequate, as the limit of liability is the sum insured. The sum insured is normally twice the estimated annual turnover. As losses paid during the period of insurance reduce the sum insured it must be reinstated and a suitable additional premium paid. Insurances can be arranged for specific contracts.

Chapter 13

Insurance of works in or extensions to existing structures by the employer: clause 22C

13.1 Clause 22C. Insurance of existing structures – insurance of Works in or extensions to existing structures [m]

22C.1 The Employer shall take out and maintain a Joint Names Policy in respect of the existing structures (which shall include from the relevant date any relevant part to which clause 18.1.3 refers) together with the contents thereof owned by him or for which he is responsible, for the full cost of reinstatement, repair or replacement of loss or damage due to one or more of the Specified Perils [o.2] up to and including the date of issue of the certificate of Practical Completion or up to and including the date of determination of the employment of the Contractor under clause 22C.4.3 or clause 27 or clause 28 or clause 28A (whether or not the validity of that determination is contested) whichever is the earlier. The Contractor, for himself and for all Nominated Sub-Contractors who are, pursuant to clause 22.3.1, recognised as an insured under the Joint Names Policy referred to in clause 22C.1 or clause 22C.3 shall authorise the insurers to pay all monies from such insurance in respect of loss or damage to the Employer.[o.2A]

22C.2 The Employer shall take out and maintain a Joint Names Policy for All Risks Insurance for cover no less than that defined in clause 22.2 [n] [0.2] for the full reinstatement value of the Works (plus the percentage, if any, to cover professional fees stated in the Appendix) and shall (subject to clause 18.1.3) maintain such Joint Names Policy up to and including the date of issue of the certificate of Practical Completion or up to and including the date of determination of the employment of the Contractor under clause 22C.4.3 or clause 27 or clause 28 or clause 28A (whether or not the validity of that determination is contested) whichever is the earlier.

The drafters of the 1986 Amendment appreciated that the insurance market may not give full all risks insurance on the existing structure and contents since they were not part of the works, and it was decided to confine the insurance obligation concerning them to the old clause 22 perils, now called specified perils. Generally speaking, while the contractor may be in a better position to insure the works, the existing structure is usually already insured by the employer but not for all risks, so it is appropriate to put the onus of insurance on the employer. Incidentally there is in any event a duty on the employer to notify the insurer of the existing structures of the proposed work.

A footnote [o.2] to both 22C.1 and 22C.2 draws attention to the fact that in some cases it may not be possible for insurance to be taken out against certain of the specified perils or the risks covered by the definition of 'all risks insurance'. In this event the definitions concerned in clauses 1.3 and 22.2 must be amended accordingly. There are two reasons for this footnote. In the first place with certain buildings, e.g. historical ones, insurance for particular risks is not available, and in some areas insurance against certain perils, e.g. flood, riot and civil commotion, may not be obtainable.

The first two sub-clauses of clause 22C treat separately the following two situations:

- Sub-clause 22C.1 deals with loss or damage *to the existing structures and the contents owned by the employer or for which he is responsible* where the insurance is against the *specified perils* (see definition in clause 1.3).
- Sub-clause 22C.2 deals with loss or damage *to the works carried out in the existing structures or for extensions thereto where the insurance is against* the risk covered by the 1986 clause 22.2 definition of *all risks insurance.*

In the *Practice Note 22 Guide to the Amendments to the Insurance and Related Liability Provisions: 1986,* issued by the JCT, it is stated that the works insurance taken out by the employer has to be in the joint names of the employer and contractor and in fact clause 1.3 now defines a joint names policy in this way. There is therefore no need for the contractor to take out insurance in respect of the works and site materials for loss or damage due to any of the risks covered by the 1986 definition of 'all risks insurance' and the previous reference to 'sole risk' in the 1980 clause 22C.1 has been omitted. Any shortfall in the insurance monies not sufficient to cover the cost of reinstatement (whether due to inadequacy of cover or to any excesses provision – uninsured amounts – in the insurance policy) is borne by the employer. However, a further question concerning the omission of the words 'shall be at the sole risk of the Employer' is whether it will have any effect on the decision in *Scottish Special Housing Association* v. *Wimpey* (1986), confirming *Archdale & Co Ltd* v. *Comservices Ltd* (1954), if that same situation arises under the new wording (see Chapter 3). Thus if the contractor negligently destroys the building by one of the insured risks, who is legally liable when clause 22C applies (or 22B for that matter)? The answer must be that the insurance arranged by the employer will operate without any subrogation rights against the contractor. Also in the case of clause 22C.1, by virtue of the last sentence, as the nominated sub-contractor (but not the domestic sub-contractor) is a joint insured, any subrogation rights against him are also waived. In effect, this House of Lords decision in the *Wimpey* case will be followed. This leaves open the position where there is a large excess under the all risks policy and prima facie the employer is liable for this amount because he has undertaken to provide insurance without excess as there is no mention of an excess in the draft policy in clause 22.2.

Probably more to the point concerning the employer's position in the

Wimpey circumstances under the current JCT wording, is the *Ossory Road* decision mentioned in Chapter 3. This is that the indemnity in clause 20.2 (that the contractor should indemnify the employer) by clause 20.3 *excepted the works* and, where applicable, clause 22C.1 (concerning the insurance of existing structures). This means that the employer again must be liable either for the works executed or for the cost of the damage to existing structures and contents as a result of a fire negligently caused by the contractor, or even so caused by the sub-contractor according to the *Ossory Road* case.

Incidentally there is a model in Appendix D of the *Practice Note 22 and Guide* referred to above to change the liability for payment of excesses under the all risks insurance which the employer is required to take out under clause 22C.2.

The employer might well consider adopting this new clause, which would be numbered clause 22C.4.4.3. It provides for the contractor to bear the cost of rectifying loss or damage which is uninsured because of excess provisions in the joint names policy to be taken out by the employer and where the loss or damage is due to negligence, etc. of the contractor or any sub-contractor. The employer has to inform the contractor at the tender stage of the amount of the excesses in the joint names policy and if they are for large amounts the employer should remember that the contractor and sub-contractor would themselves have to take out (or consider doing so) insurance for any liability for the potential uninsured loss or damage arising from the excess and this could add considerably to the cost of the works, as well as making administration more involved.

In the case of joint names policies the contractor as one of the insured must authorise the insurer to pay all monies to the employer as the other insured, and this authorisation must also be given on behalf of all nominated sub-contractors where they are joint insureds (see last sentence in clause 22C.1).

Footnote [o.2A] is mentioned at the end of clause 22C.1 and it reads as follows:

> Some Employers e.g. tenants may not be able to fulfil the obligations in clause 22C.1. If so clause 22C.1 should be amended accordingly.

22C.3 The Employer shall, as and when reasonably required to do so by the Contractor, produce documentary evidence and receipts showing that the Joint Names Policy required under clause 22C.1 or clause 22C.2 has been taken out and is being maintained. If the Employer defaults in taking out or in maintaining the Joint Names Policy required under clause 22C.1 the Contractor may himself take out and maintain a Joint Names Policy against any risk in respect of which the default shall have occurred and for that purpose shall have such right of entry and inspection as may be required to make a survey and inventory of the existing structures and the relevant contents. If the Employer defaults in taking out or in maintaining the Joint Names Policy required under clause 22C.2 the Contractor may take out and maintain a Joint Names Policy against any risk in respect of which the default shall have occurred. A sum or sums equivalent to the premiums paid or payable by the Contractor pursuant to clause 22C.3 shall be added to the Contract Sum.

It should be remembered that the insurances are different for the existing structures and contents on the one hand and the works and site materials on the other; consequently in the former case, where there is default, the contractor is given a right of entry and inspection to make a survey and inventory of the existing structures and relevant contents in the subject matter of the insurance, in order to get the sum insured correct.

In the local authorities' edition the contractor is not entitled to proof of insurance or right to insure in case of default and thus 1986 clause 22C.3 is marked 'Number not used'.

22C.4 If any loss or damage affecting work executed or any part thereof or any Site Materials is occasioned by any one or more of the risks covered by the Joint Names Policy referred to in clause 22C.2 or clause 22C.3 then, upon discovering the said loss or damage, the Contractor shall forthwith give notice in writing both to the Architect and to the Employer of the extent, nature and location thereof and

22C.4.1 the occurrence of such loss or damage shall be disregarded in computing any amounts payable to the Contractor under or by virtue of this Contract;

22C.4.2 the Contractor, for himself and for all Nominated and Domestic Sub-Contractors who are, pursuant to clause 22.3, recognised as an insured under the Joint Names Policy referred to in clause 22C.2 or clause 22C.3, shall authorise the insurers to pay all monies from such insurance in respect of the loss or damage referred to in clause 22C.4 to the Employer;

22C.4.3.1 if it is just and equitable so to do the employment of the Contractor under this Contract may within 28 days of the occurrence of such loss or damage be determined at the option of either party by notice by registered post or recorded delivery from either party to the other. Within 7 days of receiving such a notice (but not thereafter) either party may give to the other a written request to concur in the appointment of an Arbitrator under clause 41 in order that it may be determined whether such determination will be just and equitable;

.3.2 upon the giving or receiving by the Employer of such a notice of determination or, where a reference to arbitration is made as aforesaid, upon the Arbitrator upholding the notice of determination, the provisions of clauses 28A.4 and 28A.5 (except clause 28A.5.5) shall apply.

22C.4.4 If no notice of determination is served under clause 22C.4.3.1, or, where a reference to arbitration is made as aforesaid, if the Arbitrator decides against the notice of determination, then

.4.1 after any inspection required by the insurers in respect of a claim under the Joint Names Policy referred to in clause 22C.2 or clause 22C.3 has been completed, the Contractor with due diligence shall restore such work damaged, replace or repair any such Site Materials which have been lost or damaged, remove and dispose of any debris and proceed with the carrying out and completion of the Works; and

.4.2 the restoration, replacement or repair of such loss or damage and (when required) the removal and disposal of debris shall be treated as if they were a Variation required by an instruction of the Architect under clause 13.2.

As in the case of clauses 22A and 22B the last sub-clauses in clause 22C deal with insurance claims. Sub-clause 22C.4 requires the same notice in writing of loss or damage to the works, etc. by the contractor to the architect and the employer as was required in sub-clause 22B.3.1, otherwise neither the architect nor the employer would know of it. Sub-clauses 22C.4.1 and 22C.4.2 are the same as sub-clauses 22B.3.2 and 22B.3.4, respectively.

Because of the wider effect of any loss or damage (it includes the existing structures and their contents) in addition to the works, all the 22C sub-clauses are more involved than those in 22B. This is particularly so under the option of determination in sub-clause 22C.4.3.1, partly because it only applies where there is loss or damage to the works. Where equitable (within 28 days of such loss or damage) the contract may be determined by either party by registered post or recorded delivery. If objected to within seven days, written notice of reference to arbitration may be given. The words '(but not thereafter)' emphasise that the party objecting must act *at once*. Incidentally, Article 5 has been altered so that reference to an arbitrator, to decide whether a determination of the contractor's employment under clause 22C.4.3.1 is just and equitable, can be made before practical completion.

Upon determination of the contract by either party, clauses 28A.4 (retentions) and clause 28A.5 (preparation of an account) apply, except clause 28A.5.5, which obliges the employer to pay the contractor any direct loss and/or damage caused to the contractor by the determination. Clause 28A.3 states that the contractor must with reasonable despatch remove all his temporary buildings, plants, tools, equipment, goods and materials, and clause 28A.5 sets out the terms under which the contractor is to be paid. If the contract is not determined, the contractor must proceed with the work of making good as a variation required by the architect/supervising officer (see sub-clause 22C.4.4.2).

Incidentally, it may seem strange to the insurance official to find loss of profit considered as a direct loss, but the explanation lies in the decisions of the line of cases from *Millar's Machinery Co Ltd* v. *David Way & Son (1934)*, *Saint Line Ltd* v. *Richardsons, Westgarth & Co Ltd* (1940), *Wraight Ltd* v. *PH and T (Holdings) Ltd* (1968) to *Croudace Construction Ltd* v. *Cawoods Concrete Products Ltd* (1978).

The *Saint Line* case is a good example of the legal meaning of 'direct' on the one hand and 'indirect and consequential' on the other. The facts are as follows.

By a clause in a contract for the provision of engines for a ship limiting the liability of the engine builder in certain events it was provided (*inter alia*) 'nor shall their liability ever or in any case ... extend to any indirect or consequential damages or claims whatsoever'. The contract was broken by the engine builder, and the shipowner claimed damages:

- for loss of profit during the time it was deprived of the use of the vessel;
- expenses of wages, stores, etc;
- fees paid to experts for superintendence.

It was held that, inasmuch as the damages claimed were recoverable only in so far as they were the direct and natural result of the breaches, the plaintiff's rights were not cut down by the clause in the contract, since all these heads of damage were direct and immediate, and not 'indirect or consequential'.

If the reinstatement of the structure other than the works is considerable or involves work of which the contractor has no experience, the contractor or employer may wish to determine the contract. The employer may even not want to proceed with the works if the main structure is largely destroyed. Assuming there is some damage to the works, either party could carry out his wishes in accordance with this clause.

Sub-clause 22C.4.4 deals with the situation where no notice of determination is made or the arbitrator decides against the notice of determination. In this event, after any inspection required by the insurers under the joint names policy, the contractor must proceed with the work of making good as a variation required by the architect as mentioned above.

13.2 General points

Theory and practice

As in the case of clauses 22A and 22B, once again practice seems to differ significantly from wording.

Provided no notice of determination is made or the arbitrator decides against the notice of determination, then the intention is that the contractor should proceed with the repair to the works which should be treated as if it were a variation required under clause 13.2. As there is no provision for damage to the existing structures other than to require any payments to be made to the employer, any repair to the existing structure requires separate consideration. This would be the subject of a separate contract or, depending on the extent of the damage, possibly an instruction for a variation under clause 13.2 of the existing contract might suffice. The contractor might reasonably object to the latter suggestion.

Once again, it is the intention that the repairs to the damaged works should be valued by the quantity surveyor in accordance with the normal contract arrangements and the contractor should be paid through the contract, again leaving the employer to recover from the joint names insurers. In practice, however, it is usually the case that again the contractor is left to deal directly with the joint names insurers or, more likely, their loss adjusters. Once again, although payment is not usually made direct from the joint names insurers to the contractors, it is the insurers or the loss adjusters who negotiate agreement on quantum instead of the quantity surveyor doing what the contract conditions say should be done.

The model clauses

Model clauses enabling the Employer to take the 'risk' or 'sole risk' or loss or damage to the works and to the existing structure, without having to insure, are set out in *Practice Note 22* as mentioned at the end of Chapter 11.

Where the model clauses are used, the repair must still be treated as if it were a variation required under clause 13.2.

Incidentally, in the commentary in Appendix 6, the reference to clauses 22J and 22K in relation to all risks insurance is incorrect, as the insurance for the model clauses which refer to existing structures and contents can only apply to specified perils risks.

The treatment of repair

As explained in this and previous chapters the treatment of repair following loss or damage to the works is intended to be fundamentally different under clause 22A from the position under clauses 22B or 22C.

Where clause 22A applies, the contractor has no entitlement to payment under the contract for the damage repairs but instead has a contractual obligation to effect the repairs as quickly as possible and proceed with the carrying out of the completion of the works. He then relies on a successful and full recovery of outlays from the joint names insurers, under the policy which he has arranged.

In contrast, both clauses 22B and 22C give the contractor an entitlement under the contract to payment for the damage repairs whilst, not in practice but certainly in theory, it is the employer who is left with the task of effecting the successful insurance recovery. This is because the employer has arranged the cover.

This difference in treatment can result in the illogical position that the same damage involving the same contractor at the same site could produce entirely different payment results depending on whether the contract was governed by clause 22A or either clauses 22B or 22C.

As already explained, damage repairs under clauses 22B or 22C should be treated as clause 13.2 variations. That being the case, it then becomes the responsibility of the quantity surveyor to value the clause 13.2 variation (see clause 13.4.1) in accordance with provisions of the usual valuation rule (see clause 13.5). Basically, where the work is of a similar nature to that originally tendered for and executed under similar circumstances, then the bill rates should apply but, where the work is not of a similar nature not executed under similar circumstances, then daywork rates apply.

Of utmost importance is the fact that it is the quantity surveyor who has the responsibility of valuing the repair work in accordance with the provisions of the contract and once the valuation has been agreed, the Architect should issue a certificate which, in turn, should bring about payment by the employer of the valuation amount (less retention).

However, where clause 22A applies, the valuation rules have no part to play in the exercise and instead the contractor has to complete the repairs and then agree suitable indemnity with his joint names insurers or their loss adjusters. Practice shows that many differences of opinion can arise on quantum with the contractor's entitlement to profit being a most contentious issue. Quantum negotiations can often result in the contractor receiving payment of a sum different and usually considerably less than what would be paid under the contract if valued in accordance with the valuation rules.

Even more illogical, though, is the fact that the different treatment has a knock-on effect when other parts of the contract conditions are considered. A variation under clause 13.2 gives the contractor an automatic right to an extension of time under clause 25 since compliance with the architect's instructions under clause 13.2 is a relevant event (see clause 25.4.5.1). The contractor also has an entitlement to have a claim considered under clause 26 for loss and expense and architect's instruction under clause 13.2 is one of the list of matters specified (see clause 26.2.7).

Where clauses 22B and 22C apply, the fact that the damage repairs are required to be treated as a clause 13.2 variation gives the contractor the entitlement to extension of time and consideration of a claim for delay, loss and expense if of course the damage giving rise to the clause 13.2 variation is by any of the perils covered by the all risks insurance.

In contrast, the position under clause 22A is that, under clause 25, the contractor only has an entitlement to an extension of time where the loss or damage is by one or more of the specified perils (see clause 25.4.3). Furthermore, there is no entitlement to consideration of a claim for delay, loss and expense under clause 26.

It seems that these differences arise because in clauses 22B and 22C the employer is dealing with his insurer. Thus it is logical for the insurer to pay the employer. On the other hand, in clause 22A the contractor is dealing with his insurer, so there seems no reason why the money should go to the employer. Nevertheless clause 22A.4.4 appears to require payment to the employer to enable the employer to deduct professional fees before paying the contractor. Also by paying under interim certificates the employer obtains a cash flow benefit. Similar inconsistencies arise where the model clauses are used.

Chapter 14

Insurance for employer's loss of liquidated damages and date of possession, completion and postponement: clause 22D and clause 23

14.1 Clause 22D. Insurance for employer's loss of liquidated damages

22D.1 Where it is stated in the Appendix that the insurance to which clause 22D refers may be required by the Employer then forthwith after the Contract has been entered into the Architect shall either inform the Contractor that no such insurance is required or instruct the Contractor to obtain a quotation for such insurance. This quotation shall be for an insurance on an agreed value basis [o.3] to be taken out and maintained by the Contractor until the date of Practical Completion and which will provide for payment to the Employer of a sum calculated by reference to clause 22D.3 in the event of loss or damage to the Works, work executed, Site Materials, temporary buildings, plant and equipment for use in connection with and on or adjacent to the Works by any one or more of the Specified Perils and which loss or damage results in the Architect giving an extension of time under clause 25.3 in respect of the Relevant Event in clause 25.4.3. The Architect shall obtain from the Employer any information which the Contractor reasonably requires to obtain such quotation. The Contractor shall send to the Architect as soon as practicable the quotation which he has obtained and the Architect shall thereafter instruct the Contractor whether or not the Employer wishes the Contractor to accept that quotation and such instruction shall not be unreasonably withheld or delayed. If the Contractor is instructed to accept the quotation the Contractor shall forthwith take out and maintain the relevant policy and send it to the Architect for deposit with the Employer, together with the premium receipt therefor and also any relevant endorsement or endorsements thereof and the premium receipts therefor.

22D.2 The sum insured by the relevant policy shall be a sum calculated at the rate stated in the Appendix as liquidated and ascertained damages for the period of time stated in the Appendix.

22D.3 Payment in respect of this insurance shall be calculated at the rate referred to in clause 22D.2 (or any revised rate produced by the application of clause 18.1.4) for the period of any extension of time finally given by the Architect as referred to in clause 22D.1 or for the period of time stated in the Appendix, whichever is the less.

22D.4 The amounts expended by the Contractor to take out and maintain the insurance referred to in clause 22D.1 shall be added to the Contract Sum. If the Contractor defaults in taking out or in maintaining the insurance referred to in clause 22D.1 the Employer may himself insure against any risk in respect of which the default shall have occurred.

The background

If the occurrence of the risks referred to in 1986 clause 25.4.3 (specified perils) give rise to an extension of time and the fixing of a later completion date, the employer is unable to benefit from clause 24 (damages for non-completion) and consequently has no right to receive liquidated damages for the weeks or other period by which the completion date has been extended. This 1986 clause 22D is designed to give the employer the opportunity to require the contractor to arrange insurance to compensate the employer for the damages which he cannot recover under clause 24, and thus overcomes an unsatisfactory situation for the employer, particularly if delay was due to the contractor. Before the introduction of clause 22D employers with some experience of insurance, and the aid of a broker versed in the insurances required by the construction industry, would often take out an advance profits cover to protect them against the financial losses which they might incur if damage to the works during construction caused a delay in their ability to obtain rent or sell the completed building on time. It was from this situation that the Joint Contracts Tribunal decided that it was necessary to put an optional requirement into the Standard Form for the eventuality described above.

Some insurers do not like 'agreed value' sums insured but really the figure concerned can work to either party's disadvantage. Nevertheless the adoption of an agreed value does avoid any dispute over the amount of the payment due under the insurance once the policy is issued. (A copy of the Trinity Insurance Company's policy for clause 22D appears in Appendix 8).

It should be noted that the Trinity Insurance Company Ltd are no longer trading but it is understood that the Avon Insurance plc will transact this class of business. More recently the Builders' Accident will provide this cover as an extension to their contract works policy.

A footnote in the contract makes the legal point that the insurers will normally reserve the right to be satisfied that the sum referred to in clause 22D.2 is not more than a genuine pre-estimate of the damage which the employer considers, at the time he enters into the contract, he will suffer as a result of any delay. This genuine pre-estimate distinguishes the sum from a penalty clause which legally is disregarded for all purposes.

Detailed consideration

Clause 22D.1 provides for an insurance to be arranged by the contractor for the employer's loss of liquidated damages. It provides that where the

Appendix states that this insurance may be required, the architect makes the decision to inform the contractor either that no insurance is required or that he should obtain a quotation for such insurance. It is on an agreed value basis and operates until practical completion. It should be appreciated that this insurance stems from the extension of time allowed by clause 25.3 arising from delay caused by loss or damage from the specified perils to the works, work executed, site materials, temporary buildings, plant or equipment i.e. the works, etc. specified. The insurance is in respect of specified perils only and the sum insured is the rate of liquidated damages over the period selected.

Payment is based on the rate of liquidated damages calculated for the period of extension finally granted arising out of the delay caused by the specified perils (see clause 22D.3). This claims payment is subject to a qualification reading 'or any revised rate produced by the application of clause 18.1.4'. Clause 18 deals with partial possession by the employer. Clause 18.1.4 states that if part of the works is taken over by the employer, the value insured by the material damage insurance arranged by the contractor is reduced by the value of that part. This formula is carried through to the liquidated damages by clause 22D.3.

It is interesting to note that clause 22D includes cover for the consequences of loss or damage to temporary buildings, plant or equipment as these items are not specifically required to be insured elsewhere in the contract. Also elsewhere the works are required to be insured against all risks. Evidently the reluctance of insurers to offer the cover required by this clause forced this compromise.

Under clause 22D.4 the premiums paid by the contractor are to be added to the contract sum, and if the contractor defaults in providing the insurance, the employer may insure against any risks in respect of which the default shall have occurred.

Advance profits insurance

Clause 22D was provided to cope with a particular situation. Some members of the Joint Contracts Tribunal thought it was wrong that the contractor should be given an extension of time when the 'Clause 22 perils' (in the 1980 Edition) were caused by his negligence. However, the contractors' representatives, on the JCT, would not agree to exclude an extension of time when the contractor was negligent, so clause 22D became the resultant compromise.

Now the clause 22D insurance may not provide a real measure of the employer's probable loss, and it is possibly better for the employer to take out advance profits insurance, or at least to consider that option. The latter insurance can provide all risks cover, which is wider than the specified perils under clause 22D, and the employer is more likely to assess correctly the amount of his annual earnings and the effect an accident would have on them. For example, he should consider loss of rent, loss of other income, increased bank charges, and increased cost of working, if the completion date is delayed because of damage caused by an accident. In fact the Trinity Insurance Company Ltd (now not trading) compares the clause 22D insurance with advance profits insurance in the following tabulated form. Incidentally, in this table reference is made to 'Advanced Profits' insurance although, as the term relates to an interruption insurance arranged in advance of the insured's business commencing, it is arguable that 'Advance Profits' is the more correct term. Finally, in reading the following form it should be appreciated that, as these insurances are alternatives, the employer will have to come to a decision concerning advance profits insurance before instructing the contractor to obtain a quotation in accordance with sub-clause 22D.1.

Employer's loss of liquidated damages – clause 22D

The case for ...	**But, on the other hand ...**
1. Payments are for an 'Agreed Amount' per week (usually based on the anticipated rental or profit to be derived from the project) therefore the principal can budget for the amount to be received in the event of delay.	1. The 'Agreed Amount' is fixed at the outset. During a long project changes in interest rates or higher rents could make the agreed amount inadequate. (Sums insured can be amended at any time under an 'advanced profits' insurance. Also, as the policy is not 'agreed value' the premium is usually lower.)
2. Straightforward policy wording – tied in exactly with the contract conditions.	2. Cover is restricted to specified perils, as stated in the contract. (A wider range of perils cover is available under an advanced profits policy.)
3. Separate policy is issued for each project.	3. A single 'open declaration' policy can be issued for advanced profits, with charge notes for each project. 'One-off' policies are, of course, available.
4. Quotations are obtained and cover is arranged by the contractor.	4. A contractor's poor loss record could adversely affect the premium. (Under advanced profits, control of the employer's insurance arrangements remains with the employer.)

5. For contracts subject to 'sectional' completion, amounts of cover are agreed for each section at the outset.

5. The '22D' policy will only pay for damages relating to an affected section. If damage occurs to a section to be occupied by a key tenant such as Sainsbury's or Marks & Spencer, other potential tenants may be reluctant to take possession, thereby causing loss to the employer. (Advanced profits could cater for this possibility by providing cover for the development as a whole.)

6. The '22D' policy is normally subject to a two week "excess".

6. A lower 'excess' can normally be negotiated under an advanced profits policy.

Generally the decision regarding advance profits will need to be:

- whether the cover can be limited to fire and other normal additional perils; or
- whether the cover should follow the CAR insurance of the contractor, assuming that cover applies to the premises in course of erection.

The protection to be arranged depends on the type of loss likely to be incurred, which in turn depends on the intended use of the new building. If it is to be let it will be loss of rent. If it is to be used in an existing business it will be loss of profit or income.

In Appendix 9 an example is given of a specification of an advance profits policy arranged for a property developer as used by the Trinity Insurance Company. This covers loss of rent after damage covered by a CAR policy. The insured employer will have to decide how long he wishes the protection to apply, which is called the indemnity period. This period starts to run on the date on which, but for the damages to the works, the building would have been ready for occupation. The cover then follows the pattern of the normal business interruption policy.

14.2 Clause 23. Date of possession, completion and postponement – use by employer

Sub-clauses 23.1 and 23.2 refer to the date the contractor is given possession of the site, when he is expected to proceed with the works and complete them before the completion date, and also give the architect a right to postpone work. The 1986 amendment adds three new paragraphs as follows:

23.3.1 For the purposes of the Works insurances the Contractor shall retain possession of the site and the Works up to and including the date of issue of the certificate of Practical Completion, and, subject to clause 18, the Employer shall not be entitled to take possession of any part or parts of the Works until that date.

23.3.2 Notwithstanding the provisions of clause 23.3.1 the Employer may, with the consent in writing of the Contractor, use or occupy the site or the Works or part thereof whether for the purposes of storage of his goods or otherwise before the date of issue of the certificate of Practical Completion by the Architect. Before the Contractor shall give his consent to such use or occupation the Contractor or the Employer shall notify the insurers under clause 22A or clause 22B or clause 22C.2 to .4 whichever may be applicable and obtain confirmation that such use or occupation will not prejudice the insurance. Subject to such confirmation the consent of the Contractor shall not be unreasonably withheld.

23.3.3 Where clause 22A.2 or clause 22A.3 applies and the insurers in giving the confirmation referred to in clause 23.3.2 ;have made it a condition of such confirmation that an additional premium is required the Contractor shall notify the Employer of the amount of the additional premium. If the Employer continues to require use or occupation under clause 23.3.2 the additional premium required shall be added to the Contract Sum and the Contractor shall provide the Employer, if so requested, with the additional premium receipt therefor.

Clause 23.3.1 refers to the operation of clause 18 (partial possession by the employer) as an exception to the contractor's right to remain in possession of the site. While it may be that the certificate of practical completion will certify such completion at a date prior to the date of issue of the certificate, nevertheless, for the purposes of the contractor giving up possession and the cessation of the works insurance cover, the date of issue of the certificate is the only operative date.

Clause 23.3.2 caters for the employer who wishes to use or occupy the works for storage or otherwise before the date of issue of the certificate of practical completion, subject to the consent of the contractor (which cannot be unreasonably withheld), and confirmation that the insurance will not be prejudiced by such use or occupation. This occupation will be a change of risk and notification is to do with a disclosure of material fact. This operates whether clause 22A, clause 22B, or clause 22C.2 to 4 applies.

If the insurer requires an additional premium for this facility, clause 23.3.3 provides that the employer, if he still wishes to so use or occupy the works, reimburse the contractor for the additional premium if clause 22A applies.

As most all risks policies contain an exclusion making cover void from date of breach, if the employer has use or occupation of the works, observance of the above clause 23.3.2 is imperative to avoid this exclusion applying.

Chapter 15

Sub-contractors: JCT 80 and NSC/C

The Standard Form of Building Contract 1980 Edition provides for the engagement of two kinds of sub-contractor, nominated and domestic.

15.1 Introduction to the nominated sub-contract procedure

Clause 35 of JCT 80 governs the position of nominated sub-contractors. A nominated sub-contractor is defined as a sub-contractor whose selection is reserved to the architect. Such reservation can be effected by either the use of a prime cost sum or by agreement or being named in the bill of quantities, or in an instruction regarding a variation or expenditure of a provisional sum.

Prior to Amendment 10 issued in March 1991 the basic method by which a nomination was made by the architect was by the use of the 'JCT Standard Form of Nominated Sub-Contract Tender and Agreement' (Tender NSC/1), which involved the employer and nominated sub-contractor entering into the JCT Standard Form of Employer/Nominated Sub-Contractor Agreement (Agreement NSC/2). The architect issued his instruction making the nomination on the JCT Standard Form of Nomination (Nomination NSC/3). The contractor and proposed sub-contractor were then bound by a sub-contract in accordance with the terms of Schedule 2 of Tender NSC/1, one of which was that the contractor and sub-contractor comply with the general conditions of the JCT Nominated Sub-Contract NSC/4. This method was used where the sub-contract works were significant. In the alternative method, standard forms were provided, but optional with the exception of the sub-contract. They were:

- NSC/1a: form of tender.
- NSC/2a: employer/sub-contractor agreement.
- NSC/3a; nomination.
- NSC/4a: form of sub-contract.

This was a quick method where the sub-contract works were straightforward.

Amendment 10 is explained in the introduction to the *Revised procedure for Nomination of a Sub-Contractor* (published by RIBA publications) as resulting from the request of a number of constituent bodies of the JCT for simplification of the procedure. Thus the 'basic method' and the 'alternative method' were replaced by a single method which (as this publication puts it) 'avoids the to-ing and fro-ing of the Tender NSC/1 required by the basic method'.

This single method produced a new set of documents, which are mandatory whenever a nominated sub-contractor is appointed, as follows:

- NSC/T: form of tender in three parts
  (1) the invitation to tender issued by the architect to sub-contractor;
  (2) the form of tender to be submitted by each sub-contractor tendering;
  (3) the particular conditions to be agreed by contractors and sub-contractors prior to their entering into a sub-contract.
- Agreement NSC/A: the nominated sub-contract agreement to be entered into by contractors and sub-contractors subsequent to the latter's nomination by the architect.
- Conditions NSC/C: the conditions of sub-contract (derived from the old NSC/4) applicable to a sub-contract entered into on Agreement NSC/A which conditions are incorporated by reference in such Agreement.
- Agreement NSC/W: the employer/nominated sub-contractor agreement entered into between the employer and the nominated sub-contractor (derived from the old agreement NSC/2).
- Nomination NSC/N: the form it is essential the architect uses when he wishes to nominate a sub-contractor in accordance with the main contract conditions.

Points to note

Sub-contractor directly responsible to employer in certain circumstances

Agreement NSC/W places on the sub-contractor a responsibility directly to the employer for special services, such as the design of the sub-contract works, including the selection of materials and goods, to the extent that the sub-contractor has designed these works or selected goods and materials therefor. The contractor is apparently relieved of responsibility for design of the main sub-contract works (see clause 35.21 of the main contract).

Circumstances in which direct payment of sub-contractor may be enforced

Provision is also made in clause 35.13 for the architect to check, before issuing any interim certificate (other than the first) and the final certificate, whether amounts included in previous interim certificates for nominated sub-contractors have been properly discharged by the contractor. The contractor has to supply the architect with evidence of such discharge, and the sub-contractor is obliged to assist the contractor to do so. If the contractor has not discharged amounts included in previous interim certificates, the architect is required so to certify. Under the NSC/W agreement, the employer is then bound to reduce the monies he would otherwise have to pay to the contractor by the amounts undischarged and pay these amounts directly to the sub-contractor. Alternatively the sub-contractor has a direct right under NSC/W to enforce the employer to operate the provisions for direct payment following a default in payment by the contractor as certified by the architect.

Sub-contractor liable for own default

In *Westminster City Council* v. *J. Jarvis & Sons Ltd* (1970) the House of Lords criticised the fact that under the pre-1980 JCT contract, a sub-contractor could benefit from his own default. This earlier contract by sub-clause 23(g) allowed an extension of time to the main contractor because of delay on the part of the nominated sub-contractor who was not held responsible to the employer because there was no contract between the sub-contractor and the employer. A sub-contract warranty was introduced to form an agreement between the employer and nominated sub-contractor, but it was frequently not used, or was void because it was entered into after the sub-contractor had been nominated and was thus without consideration in English law. This situation has been remedied by the NSC/W provisions.

Defects and re-nomination

Provision is also made under Agreement NSC/W for the employer first to proceed against a nominated sub-contractor in respect of failure to remedy any default, etc. in the sub-contract works which appears before the issue of the final certificate. However, the contractor remains ultimately responsible for defects in materials and goods or workmanship (to the same extent that he is so responsible for such matters where he has supplied the materials or goods or the workmanship) if the employer is unable to obtain recovery from the nominated sub-contractor. The reason for this is that the contractor is responsible under the contract for the materials, goods and workmanship in the main contract works, including those parts executed by a nominated sub-contractor.

Where a nominated sub-contractor does not, for various reasons, carry out the nominated sub-contract, the architect is bound to re-nominate and the difference in cost between the original sub-contract price and the new sub-contract price is the responsibility of the employer (see main clause 35.24). This is in accordance with the decision in *North West Metropolitan Regional Hospital Board* v. *T.A. Bickerton & Son* (1970). The exception occurs where a nominated sub-contractor validly determines his employment under the sub-contract for some act or default of the contractor. Although the architect is still required to re-nominate the cost of the re-nomination falls on the contractor.

Nominated Sub-Contract NSC/C Section 6: Injury, Damage and Insurance

Definitions

Clause 6.1 states that in clauses 6.1 to 6.11 and some other clauses the phrase in column 1 shall have the meaning set out in column 2:

Column 1	Column 2
the Contractor or any person for whom the Contractor is responsible	the Contractor, his servants or agents or any person employed or engaged upon or in connection with the Works or any part thereof his servants or agents (other than the Sub-Contractor or any person for whom the Sub-Contractor is responsible), or any other person who may properly be on the site upon or in connection with the Works or any part thereof, his servants or agents; but such person shall not include the Employer or any person employed, engaged or authorised by him or by any local authority or statutory undertaker executing work solely in pursuance of its statutory rights or obligations;
the Sub-Contractor or any person for whom the Sub-Contractor is responsible	the Sub-Contractor, his servants or agents, or any person employed or engaged by the Sub-Contractor upon or in connection with the Sub-Contract Works or any part thereof, his servants or agents or any other person who may properly be on the site upon or in connection with the Sub-Contract Works or any part thereof, his servants or agents; but such person shall not include the Contractor or any person for whom the Contractor is responsible nor the Employer or any person employed, engaged or authorised by him or by any local authority or statutory undertaker executing work soley in pursuance of its statutory rights or obligations.

Injury to persons and property – indemnity to Contractor (6.2 to 6.4)

Clause 6.2 follows the wording of clause 20.1 of the main contract almost exactly subject to 'Sub-Contractor' replacing the word 'Contractor', 'Contractor' replacing the word 'Employer' and 'Sub-contract Works' for 'Works' plus the final phrase reading 'or of any local authority or statutory undertaker executing work solely in pursuance of its statutory rights or obligations'. Similarly clause 6.3 follows the wording of clause 20.2 of the main contract with the appropriate amendments. Also clause 6.4 follows clause 20.3 of the main contract to the extent that the indemnity of the sub-contractor to the contractor does not include in clause 6.3 liability or indemnity for loss or damage to the works or to site materials caused by a specified peril occurring up to and including whichever of the following is the earlier date:

- the date of the issue of the certificate of practical completion of the sub-contract;
- the date of determination of the employment of the contractor.

Consequently, to summarise the position, clause 6.2 requires the sub-contractor to indemnify the main contractor against any expense, liability, loss, claim or proceedings in respect of personal injury or death which arises out of the carrying out of the sub-contract works except to the extent that it is due to the negligence etc. of the main contractor, or his other sub-contractors, the employer or a local authority or statutory undertaker working as in the above quotation.

In clause 6.3 the sub-contractor's indemnity to the contractor against damage to property concerns only property damage caused by his negligence, breach of statutory duty, omission or default. This indemnity by clause 6.4 does not include loss or damage to the works or site materials caused by a specified peril occurring up to and including the earlier of the terminal dates. This ensures that such loss or damage is protected by the contractor's all risks policy, as under the main contract the insurers have no subrogation rights against any sub-contractor who negligently caused the specified peril. See Chapter 11 for the specified perils.

Insurance against injury to persons or property (6.5)

Clauses 6.5.1 and 6.5.2 require the sub-contractor to take out liability insurances against claims arising out of the liabilities of the sub-contractor set out in clauses 6.2 and 6.3 for personal injury or death, or damage to real or personal property which arises out of or in the course of or is caused by the carrying out of the sub-contract works. This is done by means of an employers' liability policy and a public liability policy, which is also referred to as a third party policy (see Chapter 6).

The purpose of the indemnity to the contractor given in clauses 6.2 and 6.3 is that if a third party seeks to claim from the contractor for personal injury or death or damage to property for which the sub-contractor is liable, the sub-contractor will indemnify the contractor against such claims. Clause 6.5.2, which requires the sub-contractor to insure in respect of his liability to third parties as stated in clauses 6.2 and 6.3, makes clear that the insurance taken out in accordance with that clause must meet any such claims made against the contractor by third parties as well as claims against the sub-contractor. Public liability policies will normally provide for such an indemnity under their contractual liability cover, but sub-contractors should check that their public liability policies will comply with clause 6.5.2. If the insurance proceeds are insufficient to meet the indemnity to the contractor in clauses 6.2 and 6.3 the sub-contractor under these clauses is responsible to the contractor for the shortfall.

There is a footnote to sub-clause 6.5.1 which points out that the sub-contractor has the benefit of the main contract works insurance for loss or damage by the specified perils to the sub-contract works but not for other risks, e.g. subsidence, impact, theft or vandalism. The insurance to which clause 6.5.1 refers in respect of damage to property is a public liability policy. However, such a policy will not give cover for any property such as the sub-contract works while they are in the custody and control of the sub-contractor, because it is a material or own damage policy which gives this cover, not a liability policy like the public liability policy. Consequently the latter part of clause 6.5.1 makes this position clear. Incidentally as the sub-contractor is liable for the non-specified perils mentioned above if they cause loss or damage to the sub-contract works, he may well consider that he needs to take out a works insurance himself to provide the cover he does not get under the main contract works insurance.

In clause 6.5.2 the minimum amount of the liability limit under the public liability policy which the sub-contractor must carry is stipulated in the document NSC/T Part 3 for any one occurrence or series of occurrences arising out of one event.

Clause 6.5.3 makes provision for the excepted risks (nuclear risks and sonic waves) to apply to all the insurances mentioned in this clause, i.e. the policies provided can exclude such risks, as there is no such indemnity given by the sub-contractor.

Loss or damage to the works and to the sub-contract works (6.6 to 6C)

Clause 6.6 includes three optional provisions, namely clauses 6A, 6B, and 6C, which are used when the corresponding clauses 22A, 22B, and 22C apply in the main contract. All three of these NSC/C clauses have similar provisions concerning damage to the sub-contract works.

According to these three clauses, whoever is responsible under JCT 80 for providing a CAR policy covering the main contract works shall ensure that the policy is issued or endorsed that, in respect of damage by the specified perils, the sub-contractor is recognised as an insured under the policy, or the insurers waive any rights of subrogation they may have against the sub-contractor. This recognition or waiver must continue up to and including whichever is the earlier of the terminal dates mentioned under the previous heading concerning clauses 6.2 to 6.4. It should be appreciated that this waiver or inclusion of the sub-contractor as a joint names insured means that the sub-contractor will not be responsible for damage to his work by a specified peril even if it was caused by his own negligence. The specified perils were set out in Chapter 11.

Clause 6A.2.1 states that the sub-contractor is responsible for the cost of restoration of sub-contract work, replacement or repair of site materials for the sub-contract works and removal of any debris in accordance with clause 6A.3 *except* to the extent that such loss or damage is due to:

- one or more of the specified perils (whether or not caused by the negligence, breach of statutory duty, omission or default of the sub-contractor or any person for whom the sub-contractor is responsible); or
- any negligence, breach of statutory duty, omission or default of the contractor or any person for whom the contractor is responsible or of the employer or any person employed, engaged or authorised by him or any local authority or statutory undertaker executing work solely in pursuance of its statutory rights or obligations.

Clause 6A.2.2 only applies where non-specified perils occur and will be considered under the next heading.

Clauses 6A.3 and 6A.4 deal with the procedure when loss or damage occurs to the sub-contract works or site materials for the sub-contract works. Clause 6A.4 states that, where the sub-contractor is not responsible for the cost of restoration of sub-contract works etc. under clause 6A.2, it shall be treated as if it were a variation.

Under clause 6A.5 on or after the earlier of the terminal dates the sub-contractor shall not be responsible for loss or damage to the sub-contract works except to the extent it is caused by the negligence etc. of the sub-contractor or any person for whom the sub-contractor is responsible.

Clauses 6B and 6C and their sub-clauses follow the wording which has just been described in clause 6A except that allowance has to be made for the fact that the employer is responsible for the joint names policy provided under the main contract and under clause 6C the insurance for existing structures and their contents is dealt with separately under the main contract as is the determination of the contractor's employment by the loss or damage.

Damage caused by the non-specified perils

Clauses 6A, 6B, and 6C deal with responsibility for damage by the non-specified perils in the same way. Thus clause 6A.2.2 provides that, where during the progress of the works sub-contract materials and goods have been fully, finally and properly incorporated into the works before practical completion of the sub-contract works, the sub-contractor shall only be responsible for their loss or damage where it is caused by his own negligence etc. or that of any person for whom he is responsible. It has already been mentioned that clause 6A.2.1 provides that the sub-contractor is not responsible for such property to the extent that its loss or damage is due to the negligence etc. of the contractor, the employer, local authority or statutory undertaker.

To summarise the position:

- the sub-contractor is responsible for the damage to the sub-contract work, goods and materials *before* they are fully, finally and properly incorporated into the works, except to the extent that the loss or damage is caused either by a specified peril or by the negligence, etc. of the contractor, employer, local authority or statutory undertaker;
- the contractor is responsible for the damage to the sub-contract work, materials and goods *after* they are fully, finally and properly incorporated into the works except to the extent that the loss or damage is due to the negligence etc. of the sub-contractor or any person for whom the sub-contractor is responsible.

How is the phrase 'fully, finally and properly incorporated into the main contract works' to be interpreted? Most contractors consider that the whole unit in the case of, say, a roof or wall must be completed before the tiles or bricks involved can be considered as fully, finally and properly incorporated. It has been argued that in the case of electrical, gas and heating installations the complete installation must be connected up and tested, and the service operating before this phrase applies. On the other hand it has been said that because the introductory words of clause 6A.2.2 read:

> Where during the progress of the Sub-Contract Works, sub-contract materials or goods have been fully, finally and properly incorporated into the Works before practical completion of the Sub-Contract Works

it is fairly clear that the heating system does not need to be filled and tested before the goods and materials in it are fully, finally and properly incorporated. As there are no legal decisions on the point it is advisable for the parties concerned to agree the interpretation when the contract is negotiated.

Policies of insurance – production – payment of premiums – default by contractor or sub-contractor (6.7 to 6.10)

Clause 6.7 concerns the production by the sub-contractor of documentary evidence of insurance required under clause 6.5, to the contractor.

Clause 6.8 deals with the position if the sub-contractor defaults in insuring as provided in clause 6.5, i.e. the contractor may take out such insurance, the premium for which is paid by the sub-contractor. Clause 6.5 concerns employers' and public liability policies.

Clause 6.9 states that, except where the main contract conditions include clause 22B or clause 22C of a local authorities version of the standard form the contractor shall produce documentary evidence of the insurance of the works, etc. in accordance with clause 6A.1, clause 6B.1 or clause 6C.1, whichever is applicable, to the sub-contractor.

There is a footnote to clause 6.9 which explains that the introductory exception is included because the local authorities versions of the standard form do not provide any right for the contractor to require the employer to produce any documentary evidence or relevant policy or policies and premium receipts. Presumably this is because local authorities may be their own insurers.

Clause 6.10 deals with the position if the contractor defaults in insuring as provided in clause 6.9, i.e. the sub-contractor may take out such insurance the premium for which is paid by the contractor.

Sub-contractor's plant, etc. – responsibility of contractor (6.11)

The contractor by this clause is only responsible for loss or damage to the plant, tools, equipment or other property belonging to or provided by the sub-contractor, his servants or agents or sub-sub-contractors and to any materials or goods which are not for the sub-contract works, if the contractor or those for whom he is responsible have been negligent, etc. This means the contractor is to be indemnified for any expense he incurs following such loss or damage unless it was due to the contractor's negligence etc. Therefore it is for the sub-contractor to decide to what extent he insures the property mentioned in this clause in respect of material damage or legal liability.

15.2 Domestic sub-contractors and the DOM/1 Sub-Contract

Whereas nominated sub-contractors are those whose final selection and approval are reserved to the architect, domestic sub-contractors are defined in the JCT *Practice Note 9* as any other sub-contractors whose selection remains with the contractor.

Clause 19 of the JCT main contract governs the situation. If the contractor wishes to sub-contract any portion of the works to a domestic sub-contractor (which is called in clause 19 'sub-letting'), he must obtain the written consent of the architect to such sub-letting before he engages a sub-contractor but such consent cannot be unreasonably withheld. The Building Employers' Confederation publishes the DOM/1 sub-contract.

The clauses in the DOM/1 contract follow closely the clauses of NSC/C in dealing with third party injury damage and its insurance plus responsibility for the sub-contract works and its insurance, although the clause numbers are different. Thus while NSC/C and DOM/1 both start at clause 6 and follow exactly the sub-clause numbers for the section headed 'Injury to persons and property – indemnity to Contractor' and the wording is followed almost exactly, in subsequent clauses in DOM/1 the numbers change. However, in the subsequent clauses the wording does not alter very much except for two variations mentioned later. The numbers vary in accordance with the following section headings:

NSC/C Clauses	Clause headings	DOM/1 Clauses
6.2 to 6.4	Injury to persons and property – indemnity to Contractor	6.1 to 6.4
6.5	Insurance against injury to persons or property	7.1 to 7.3
6.6 to 6C.5	Loss or damage to the Works and to the Sub-Contract Works	8.1 to 8C.5
6A	Sub-Contract Works in new buildings – Main Contract Conditions Clause 22A	8A
6B	Sub-Contract Works in new buildings – Main Contract Conditions Clause 22B	8B
6C	Sub-Contract Works in existing structures – Main Contract Conditions Clause 22C	8C
6.7 to 6.10	Policies of insurance – production – payment of premiums	9.1 to 9.4
6.11	Sub-Contractor's plant, etc. – responsibility of Contractor	10

The first real variation with NSC/C concerning terminal dates

The first variation, mentioned earlier, is a definition of 'Terminal Dates' in clause 6.1.2 in DOM/1 which reads as follows:

2 The term 'Terminal Dates' shall mean:

either the date of the written notice of the Sub-Contractor under clause 14.1 provided the Contractor does not dissent therefrom under clause 14.1 or, where the Contractor does so dissent, the date upon which the Contractor issues in writing to the Sub-Contractor a confirmation [b] of the agreement under clause 14.2 or, failing such agreement, the date of issue of the certificate of Practical Completion of the Works under clause 17.1 of the Main Contract Conditions, whichever is applicable; or

the date of determination of the employment of the Contractor (whether or not the validity of that determination is contested) under clause 22C.4.3 (where applicable) or clause 27 or clause 28 of the Main Contract Conditions.

In the NSC/C the terminal dates were defined as whichever is the earlier of the issue of the certificate of practical completion of the sub-contract or the date of determination of the employment of the contractor. However, in DOM/1 there are various ways of proving that practical completion of the domestic sub-contract has taken place, some of which do not require the issue of a certificate, e.g. mere notice by the sub-contractor to the contractor under the terms of clause 14.1 is sufficient.

The importance of this is seen under clause 22.3.2 of the main contract, which deals with the recognition of each domestic sub-contractor as an insured under the relevant joint names policy or the waiver of subrogation rights against domestic sub-contractors and states that:

> Such recognition or waiver for Domestic Sub-Contractors shall continue up to and including the date of issue of any certificate *or other document* which states that the domestic sub-contract works are practically complete.

Consequently DOM/1 allows for practical completion in accordance with the wording of clause 6.1.2 quoted above. Therefore the contractor must agree in writing to the date of practical completion of the domestic contract works to end the cover the sub-contractor gets under the relevant joint names policy applicable by virtue of clause 22 of the main contract.

The footnote [b] to clause 6.2.1 indicates that although written confirmation of any such agreement is not expressly required under clause 14.2 of DOM/1 it should be issued on the same date that such agreement is reached.

The second real variation with NSC/C concerning work on existing structures

The second variation where the wording of DOM/1 differs from that in NSC/C occurs because *domestic* sub-contractors do *not* get recognition as joint insureds, or alternatively a waiver of subrogation rights against them, where work is being done on existing structures or extending existing structures under clause 22C of the main contract where the employer is required to insure the existing structures and their contents against the specified perils. Thus the sub-contract works will be covered by the sub-contractor's all risks policy and the main works should be covered by his public liability policy in the event of the loss or damage being caused by his negligence.

It should be noted that exception 4(b) in Section 2 (Public Liability) of the combined contractors' policy in Appendix 2 excludes property in the insured's custody or control. This could be a disaster for the insured (contractor or sub-contractor) who is legally liable for property he is working on. However, in the policy in the Appendix it will be seen that there is a proviso stating that the exception does not apply to damage to buildings (including contents) not owned or leased or rented by the insured but temporarily occupied by the insured for the purpose of maintenance, alteration, extension, installation or repair. In the case of some older policies this proviso may not appear so it is important for the sub-contractor to check that his policy provides this cover.

This second variation, which occurs when comparing the wording of DOM/1 with NSC/C, first shows itself in the alteration of the wording in sub-clauses 6.3 and 6.4, i.e. there is reference to clause 6C in NSC/C but no such reference to the equivalent clause in DOM/1 as the latter contains no reference to the work on existing structures or their contents in clause 8C.

The second variation, as just mentioned, shows itself in clause 8C by omitting any reference to loss or damage by the specified perils to the existing structures and contents thereof for the same reason, i.e. domestic sub-contractors do not get the recognition or waiver as explained above.

15.3 Incompatibility between the sub-contract and the main contract

In Practice Note 22, Part D, pages 44–46 inclusive, it states clearly that the nominated sub-contractor is responsible for the nominated sub-contract works and site materials by risk other than the specified perils (except of course for damage due to the negligence of the contractor, etc. or damage to materials fully, finally and properly incorporated into the works). Domestic sub-contractors are in a similar position as regards non-specified perils. Therefore, there is no doubt that it is the intention of the JCT to leave the risks of non-specified perils damage to the sub-contract works at the risk of the sub-contractor.

However, in the main contract clauses 22B and 22C, the employer cannot avoid contractual responsibility for all risks of physical loss or damage to the works including all works sub-contracted. Thus, the contractor has rights under the main contract against the employer but, at the same time under the sub-contract, the contractor can insist on the sub-contractor carrying out work as explained above free of charge. It is complex but in fact the employer's contractual obligation would be subject to a right to subrogate against the sub-contractor to the extent that clause 22.3 does not apply to non-specified perils damage.

An example of more direct compatibility which can be used is shown where model clauses 22F and 22H are used (see Appendix 6).

These model clauses deal with the circumstances where the employer does not wish to insure but is willing to take the 'sole risk'. Both of these model clauses deal with an all risks loss or damage to the works and both treat the repair of such loss or damage as a variation required under clause 13.2. However, both model clauses are then qualified so that the employer is not required to accept the 'sole risk' and therefore responsibility for payment for non-specified perils damage to nominated or domestic sub-contract works.

The method used is to make the clause which replaces clauses 22B and 22C subject to clause 22.3 which, in turn, replaces clauses 22.3.1 and 22.3.2. There is further qualification making the employer responsible for all loss or damage to the sub-contract works where the works, materials or goods have been fully, finally and properly incorporated into the works. Consequently, where these model clauses are used, there is a compatibility for the position under the sub-contract form and the main contract form and a similar qualification could be made to clauses 22B and 22C. This situation is not complicated by insurance and subrogation, as clauses 22B and 22C are.

Another interesting problem arises in respect of unfixed materials and

The result is that, apart from retaining works insurance for the existing list of perils only, the amendments follow certain of the changes set out in Amendment 2, November 1986, for the Standard Form 1980 Edition.

The liability (indemnity), responsibility for the works and insurance provisions of this agreement are all set out in clause 6, which is divided into four sub-clauses. The third sub-clause is divided into two optional sub-clauses. These sub-clauses are as follows:

- 6.1   Injury to or death of persons
- 6.2   Injury or damage to property
- 6.3A Insurance of the Works by Contractor – Fire, etc.
- 6.3B Insurance of the Works and existing structures by Employer – Fire, etc.
- 6.4   Evidence of insurance

Sub-clause 6.1. Injury to or death of persons

> The Contractor shall be liable for and shall indemnify the Employer against any expense, liability, loss, claim or proceedings whatsoever arising under any statute or at common law in respect of personal injury to or death of any person whomsoever arising out of or in the course of or caused by the carrying out of the Works, except to the extent that the same is due to any act or neglect of the Employer or of any person for whom the Employer is responsible. Without prejudice to his liability to indemnify the Employer the Contractor shall take out and maintain and shall cause any sub-contractor to take out and maintain insurance which, in respect of liability to employees or apprentices, shall comply with the Employer's Liability (Compulsory Insurance) Act 1969 and any statutory orders made thereunder or any amendment or re-enactment thereof and, in respect of any other liability for personal injury or death, shall be such as is necessary to cover the liability of the Contractor or, as the case may be, of such sub-contractor.

This is both a liability and an insurance clause.

The plan here is to include within the liability and indemnity to the employer the complementary insurance clause. So while the first sentence follows the wording of clause 20.1 of the JCT 80 contract it is followed immediately by the wording of clause 21.1.1 of the JCT 80 contract requiring insurance, in an abbreviated form to cater firstly for the contractor's liability to his employees and secondly for third party liability. The only difficulty is that the indemnity to the employer includes a responsibility on the part of the contractor to see that the sub-contractor takes out the necessary liability insurances.

Now in the main contract the words 'and shall cause any sub-contractor to take out and maintain' were deleted by Amendment 4 issued July 1987. However, the same thing was not done in the Minor Works form. There should be no difficulty as far as the employer's liability insurance is concerned as the statute mentioned makes it compulsory for sub-contractors to insure in this respect. In the case of the public liability aspect, however, any sub-contract must be drafted to ensure that the contractor requires the sub-contractor to provide the necessary insurance cover.

In the Minor Works form there are no provisions which deal with what is in effect the nomination of a sub-contractor nor is there any form of sub-contract which would be applicable to selected sub-contractors. Clause 3.2 allows sub-contracting subject to written consent of the architect, but in view of clause 6.1 the contractor is made responsible for seeing that his domestic sub-contractors bear their responsibility for 'Injury to or death of persons' by ensuring that they have proper insurance cover. Probably the best way to do this is for the contractor to demand evidence of insurance from the sub-contractor in the same way that it is required from the contractor in clause 6.4 (see later).

Presumably if the sub-contractor has not taken out the necessary insurance cover the contractor will be held liable for this default on the part of the sub-contractor. A second line of defence for the contractor would be to require the sub-contractor to provide a performance bond. In the first place the sub-contractor is unlikely to obtain such a bond without providing the surety with proof of the conventional insurances, including the two required under clause 6.1. Secondly it is a protection against the sub-contractor becoming insolvent.

Sub-clause 6.2: Injury or damage to property

> The Contractor shall be liable for, and shall indemnify the Employer against, any expense, liability, loss, claim or proceedings in respect of any injury or damage whatsoever to any property real or personal (other than injury or damage to the Works or to any unfixed materials and goods delivered to, placed on or adjacent to the Works and intended therefor or, where clause 6.3B is applicable, to any property insured pursuant to clause 6.3B for the perils therein listed) insofar as such injury or damage arises out of or in the course of or by reason of the carrying out of the Works and to the extent that the same is due to any negligence, breach of statutory duty, omission or default of the Contractor, his servants or agents, or of any person employed or engaged by the Contractor upon or in connection with the Works or any part thereof, his servants or agents. Without prejudice to his obligation to indemnify the Employer the Contractor shall take out and maintain and shall cause any sub-contractor to take out and maintain insurance in respect of the liability referred to above in respect of injury or damage to any property real or personal other than the Works which shall be for an amount not less than the sum stated below for any one occurrence or series of occurrences arising out of one event:
>
> insurance cover referred to above to be not less than:
>
> £. . . . .

This is both a liability and an insurance sub-clause.

In this clause the words 'property real or personal' excludes the works, unfixed materials and goods intended for the works. Also where clause 6.3B is applicable, concerning loss or damage to existing structures and their contents, such property is excluded from the words in quotes. This is because clauses 6.3A and 6.3B deal with this type of property. Consequently this clause 6.2 generally follows in the first part (the liability part) the intention of clauses 20.2 and 20.3 of the main contract in an abbreviated form.

Clause 6.2 then incorporates the insurance requirement, including the contractor's responsibility for the sub-contractor insuring as explained in clause 6.1 above. The insurance requirement contains a limit of indemnity for any one occurrence or series of occurrences arising out of one event. The insurance policy required by this clause only refers to the public liability policy because this policy covers damage to property, thus the employers' liability policy is not involved. The interesting point is that according to the contract wording the limit of indemnity only applies to damage to property claims and not to injury claims, but in practice the limit of indemnity in the public liability policy applies to personal injury claims as well as property claims. A public liability policy with unlimited liability for personal injury claims is unobtainable, as far as any one occurrence or series of occurrences arising out of one event are concerned. Public liability policies are usually unlimited for the period of insurance, i.e. yearly. Incidentally the employers' liability policy since the 1st January 1995 also has a limit, usually £10 million in respect of any one occurrence. So as in the previous sub-clause a condensed and simplified form of clause 21.1.1 of the main contract is produced here.

No provision is made in the Minor Works Agreement for the employer being responsible for new works, as will be seen later, and also there is no provision for the employer to take out (or arrange for the contractor to take out) a clause 21.2.1 insurance to protect himself against third party liability (and his own property other than the works) for the collapse risks. Therefore, it is for the employer to arrange this cover if it is considered necessary (see Chapters 9 and 10).

Throughout clause 6 no reference is made to the excepted perils of nuclear risks and sonic waves as in the main contract. Sonic waves are of little importance as such damage is unlikely to occur, and certainly not at the catastrophe level as was originally feared. However, no insurer will cover nuclear risks. As explained elsewhere in this book this risk is met by the nuclear reactor owner by statute up to a certain amount and then by the government. Nevertheless this risk should be excluded, if only for the sake of tidiness, as in the case of the main contract.

Sub-clause 6.3A: Insurance of the works by contractor [j] [j.1]

The Contractor shall in the joint names of Employer and Contractor insure the Works and all unfixed materials and goods delivered to, placed on or adjacent to the Works and intended therefor against loss and damage by fire, lightning, explosion, storm, tempest, flood, bursting or overflowing of water tanks, apparatus or pipes, earthquake, aircraft and other aerial devices or articles dropped therefrom, riot and civil commotion, for the full reinstatement value thereof plus ...% [k] to cover professional fees.

After any inspection required by the insurers in respect of a claim under the insurance mentioned in this clause 6.3A the Contractor shall with due diligence restore or replace work or materials or goods damaged and dispose of any debris and proceed with and complete the Works. The Contractor shall not be entitled to any payment in respect of work or materials or goods damaged or the disposal of any debris other than the monies received under the said insurance (less the percentage to cover professional fees) and such monies shall be paid to the Contractor under certificates of the Architect/the Contract Administrator at the periods stated in clause 4.0 hereof.

This sub-clause and sub-clause 6.3B are alternatives as explained in footnote [j]. This insurance of the works, unfixed materials and goods delivered to placed on or adjacent to the works and intended therefor, against fire and specified perils, (see Chapter 12 for details of this policy) has to be in the joint names of the employer and contractor. It has already been explained at the beginning of this chapter why all risks cover is not required. The usual percentage for professional fees is 15% to 20% (see footnote [k] which states that the percentage should be inserted).

The second paragraph of sub-clause 6.3A deals with the position after the inspection of a claim by the insurers. The contractor is to proceed with the necessary work, presumably after acceptance of the claim by the insurers. In any event the contractor has a duty under clause 1.1 to carry out and complete the works. Payment is limited to the amount of the insurance monies (less the percentage to cover professional fees).

The duration of the fire and special perils cover would be expected to run from the date of the commencement of the contract to the date of practical completion under clause 2.4 when the employer would take over responsibility for insurance (see the case of *English Industrial Estates Corporation* v. *George Wimpey & Co Ltd* (1973) in Chapter 7). The footnote [j.1] states that where the contractor has in force an all risks policy which insures the works against loss or damage by *inter alia* the perils referred to in clause 6.3A this policy may be used to provide the insurance required by clause 6.3A, provided the policy recognises the employer as a joint insured with the contractor in respect of the works and the policy is maintained.

Sub-clause 6.3B: Insurance of the works and any existing structures by Employer [j]

The Employer shall in the joint names of Employer and Contractor insure against loss or damage to any existing structures (together with any contents owned by him or for which he is responsible) and to the Works and all unfixed materials and goods delivered to, placed on or adjacent to the Works and intended therefor by fire, lightning, explosion, storm, tempest, flood, bursting or overflowing of water tanks, apparatus or pipes, earthquake, aircraft and other aerial devices or articles dropped therefrom, riot and civil commotion.

If any loss or damage as referred to in this clause occurs to the works or to any unfixed materials and goods delivered to, placed on or adjacent to the Works and intended therefor then the Architect/the Contract Administrator shall issue instructions for the reinstatement and making good of such loss or damage in accordance with clause 3.5 hereof and such instructions shall be valued under clause 3.6 hereof.

In this alternative sub-clause the employer is required to insure in the joint names of the employer and the contractor the property stated in the heading against fire and special perils.

The second paragraph deals with loss or damage as referred to in this clause when the Architect/the Contract Administrator is required to issue instructions for the reinstatement of such loss or damage in accordance with clause 3.5. This seems to mean the equivalent of a variation required by the architect, as reference is also made to clause 3.6, which is headed 'Variations', and follows in a different form the provision in clause 13 of the main contract.

Sub-clause 6.4: Evidence of insurance

The Contractor shall produce, and shall cause any sub-contractor to produce, such evidence as the Employer may reasonably require that the insurances referred to in clauses 6.1 and 6.2 and, where applicable, 6.3A hereof have been taken out and are in force at all material times. Where clause 6.3B hereof is applicable the Employer shall produce such evidence as the Contractor may reasonably require that the insurance referred to therein has been taken out and is in force at all material times.

This sub-clause deals with evidence of insurances for all the sub-clauses of clause 6 whereas the main contract deals with such evidence clause by clause. However the intention is the same, namely the party responsible for producing the insurance cover has to produce to the other party evidence of such insurance. Chapter 4 should be referred to as the matter is dealt with there in some detail.

A case illustrating the importance of identifying the actual contract concerned

National Trust v. *Haden Young* (1994) involved the revised January 1987 version of the Minor Building Works Agreement, not the 1994 form set out in this chapter. It should be appreciated that in the 1987 version clause 6.2 makes the contractor responsible for loss or damage to any property (other than damage to the works). Thus the qualification in brackets does not include, as the 1994 version does the words 'or, where clause 6.3B is applicable, to any property insured pursuant to clause 6.3B for the perils therein listed'. The effect of the omission of these words was to make the contractor responsible to the employer for damage to the existing structures and contents if it was caused by the negligence of the contractor. The omission of the words in quotes from clause 6.2 meant that the contractor remained liable for the negligence of his sub-contractor. Haden Young, in this case, in accordance with the wording of clause 6.2, which refers to the negligence of 'any person employed or engaged by the contractor'.

In this case Haden Young admitted negligence for the fire damage but argued that the contractual arrangements in force relieved it from liability. There was no written sub-contract between the contractor and Haden Young and, although the contractor was not a party to the proceedings, most of the issues between the National Trust and Haden Young depended on the true construction of the Minor Works Agreement where the National Trust was the employer.

What are the lessons to be learnt from this case? Briefly, always check the contract wording. From the contractor's point of view, the National Trust had not insured under a joint names policy as required by clause 6.3B. If they had it would have made no difference to the main contractor, who was in any event responsible for negligent fire damage to the existing structures and contents under clause 6.2, except that he could have claimed an indemnity under that policy. If the Minor Works form of 1994 had been in operation in the case under discussion the contractor would not be liable under clause 6.2, because of the difference in wording. Similarly with the Standard Form. From the sub-contractor's point of view, and the existence of a joint names policy, it should be noted that Haden Young had a difficulty in that they were not a party to the contract between the contractor and the National Trust. There had been a previous authority (*Norwich City Council* v. *Harvey and Briggs Amasco* (1989)) which decided that a sub-contractor could obtain the benefit of the joint names policy purely because of the scheme of arrangement under which the building works were carried out. The argument was that the sub-contractors have an unwritten direct contract with the employer saying that in return for them carrying out the work the employer undertakes to insure the building in respect of loss or damage by fire. The appeal court in the *National Trust* case rejected this argument by using the words of the judge of the court of first instance who said:

> 'Clause 6 when read as a whole indicates that the parties (the employer and the main contractor) expressly differentiate between the contractor and sub-contractor. In 6.1 and 6.2 there is an express reference to "cause any sub-contractor", similarly in clause 6.4. 6.3B refers to the "contractor" only. I can see no reason to extend the word "contractor" in its context as [counsel] contends. There is no convincing commercial reason why I should. I see no necessity to do so to give efficacy to the contract. I see no reason to override the express language of the parties. Thus any insurance cover taken out in the joint names of the employer and the main contractor would not have enured to the benefit of the sub-contractors.'

Consequently although Haden Young embarked on the sub-contract works on the assumption that the employer would be insuring against fire and the insurance would pay, it was disappointed. The fact that the National Trust did not take out insurance, as mentioned earlier, hardly mattered as an insurer would have had a right against any liable third party (the sub-contractor in this case).

If the 1994 version of the Minor Works Form had been used the sub-contractor would not have been relieved of responsibility because of the current wording of clause 6.2. It is only the contractor who would have been so relieved.

The words of Nourse LJ in the Court of Appeal should act as a warning to all those involved in construction contracts. He said:

> 'The appellant's arguments have demonstrated that it is quite easy to make a short and simple question of construction look long and difficult if the court allows itself to be deflected from its proper function of keeping to the words of the provisions to be construed and not working backwards from other decisions on other forms of contract...'

Finally it is worth noting that if the Standard Form JCT 80 with the 1986 amendment had applied in the Haden Young case and Haden Young had been a nominated sub-contractor, then if there had been a joint names policy Haden Young would have benefited from the cover provided by that policy (see clause 22.3). Incidentally the Minor Works Form has no standard sub-contract form for use with it.

16.2 JCT Intermediate Form, IFC 84

When this form was originally issued it was stated to be for contracts in the range between those for which the JCT Standard Form of Building Contract (1980 Edition) and the JCT Agreement for Minor Building Works (1980) were issued. This was expanded by the following statement, namely that the Form would be suitable where the proposed building works are:

- of a simple content involving the normally recognised basic trades and skills of the industry; and
- without any building service installations of a complex nature, or other specialist work of a similar nature; and
- adequately specified, or specified and billed, as appropriate prior to the invitation of tenders.

The revised insurance and liability provisions for the JCT Intermediate Form, IFC 84, are contained in Amendment 1, issued November 1986, for IFC 84. They are substantially the same as those in Amendment 2, issued November 1986, for JCT 80. IFC 84 is broadly based on JCT 80 but the clause numbers differ. The following table lists clause numbers in Amendment 1 for IFC 84 with their equivalent in Amendment 2 for JCT 80. They do not call for comment as the commentary given earlier in Chapters 4, 9, 11 and 13 to the clauses under Amendment 2 applies. Clauses not mentioned in Table 16.1 are those requiring further explanation, as follows.

Table 16.1

Amendment 1 – (IFC 84)	Amendment 2 – (JCT 80)	Clause
8.3	1.3	Definitions
6.1.1	20.1	Injury to persons
6.1.2	20.2	Damage to property
6.1.3	20.3.1	Damage to property – exclusion of the works, etc.
6.2.1	21.1.1.1 21.1.1.2	Contractor's insurance for personal injury and damage to property
6.2.2	21.1.2	Evidence of insurance
6.2.3	21.1.3	Contractor defaults in taking out insurance
6.2.4	21.2.1 to 21.2.4	Insurance – liability, etc. of employer
6.2.5	21.3	Excepted risks
6.3	22	Insurance of the works
6.3.1	22.1	Insurance of the works – alternative clauses
6.3.2	22.2	Definitions – all risks insurance – site materials
6.3A.1	22A.1	New buildings – contractor to take out all risks insurance
6.3A.2	22A.2	Single policy – insurers approved by employer – contractor fails to insure
6.3A.3.1	22A.3.1	Annual policy of contractor – alternative to 22A.2
6.3A.3.2	22A.3.2	Default in insuring – 22A.2 to apply
6.3A.4.1 to 6.3A.4.5	22A.4.1 to 22A.4.5	Loss or damage to works – insurance claims – contractor's obligations – use of insurance monies
6.3B.1	22B.1	New buildings – employer to take out all risks insurance
6.3B.3.1 to 6.3B.3.5	22B.3.1 to 22B.3.5	Loss or damage to works – insurance claims – contractor's obligations – payment by employer
6.3C.1	22C.1	Existing structures – specified perils – employer to take out joint names policy
6.3C.3	22C.3	Failure of employer to insure – rights of contractor
6.3C.4	22C.4	Insurance claims – notice by contractor to architect and employer
6.3C.4.1 to 6.3C.4.4	22C.4.1 to 22C.4.4	Contractor's obligations – payment by employer – determination of contract
6.3D.1 to 6.3D.4	22D.1 to 22D.4	Insurance for employer's loss of liquidated damages – JCT 80 clause 25.4.3, IFC clause 2.4.3
2.1	23.3.1 to 23.3.3	Possession by contractor – use or occupation by employer

While clause 6.2.1 appears in Table 16.1 it was only brought into line with the Standard Form by Amendment 7 of the IFC issued in April 1994. This Amendment (similar to Amendment 4 of JCT 80, see Chapter 4) contains the following Guidance Note:

'This amendment recognises that the previous provisions in clauses 6.2.1, 6.2.2 and 6.2.3 in relation to public liability insurance by sub-contractors are not in practice honoured; and that in a great majority of cases cannot be honoured as many small sub-contractors do not carry insurance indemnities which match those required from the Main Contractor under clause 6.2.1. This amendment therefore leaves the level of indemnity to be given by the sub-contractors in respect of public liability insurance as a matter to be settled between the Contractor and sub-contractor. The Employer remains protected by the indemnity given by the Contractor under clauses 6.1.1 and 6.1.2 and his right under clause 6.2.1 to state by entry in the Appendix the amount of insurance cover the Contractor must provide for any one occurrence or series of occurrences arising out of one event.

The purpose of the indemnity to the Employer given in clauses 6.1.1 and 6.1.2 is that if a third party seeks to claim from the Employer for personal injury or death or injury or damage to property for which the Contractor is liable the Contractor will indemnify the Employer against such claims. Clause 6.2.1 (which requires the Contractor to insure in respect of his liability to third parties as stated in clauses 6.1.1 and 6.1.2), as amended by Amendment 7, makes clear that the insurance taken out pursuant to that clause must meet any claims made against the Employer by third parties as well as claims against the Contractor. Third party (public liability) policies will normally provide for such an indemnity but Contractors should check that their third party policies will comply with the amended clause 6.2.1.

The amendment makes **no change** to the position that if the insurance proceeds are insufficient to honour the indemnity given to the Employer in clauses 6.1.1 and 6.1.2 the Contractor under these clauses is responsible for that shortfall to the Employer.'

The Appendix to Practice Note 1N/1 on IFC 84 contains an additional clause which can be adopted by the parties to incorporate the same effect as clause 18 of the main contract (partial possession by the employer). The Appendix to 1N/1 and the additional clause have been revised to follow the revised clause 18 in JCT 80, 1986 version.

IFC clause 6.3.3 is equivalent to clause 22.3 of the main contract ('Nominated and domestic sub-contractors – benefit of joint names policies – specified perils') but it distinguishes between domestic sub-contractors referred to in clause 3.2 and those referred to in clause 3.3 ('Named persons as sub-contractors'). Only the latter type of sub-contractor (like nominated sub-contractors under JCT 80 as revised) obtains the benefit, where clause 6.3C applies, of the joint names policy taken out under clause 6.3C.1 in respect of the existing structures and their contents. Both types of sub-contractor obtain the benefit, in respect of loss or damage to the works and site materials, of the joint names policies for the works under clause 6.3A or clause 6.3B or clause 6.3C.2. The benefit for sub-contractors to whom clause

3.2 applies lasts *up to and including the date of issue of any certificate or other document which states that the Sub-Contract Works . . . are practically complete* or, if earlier, until determination of the main contractor's employment. The benefit for sub-contractors named under clause 3.3 lasts *up to and including the date of issue of the certificate of practical completion of the Sub-Contract Works (as referred to in clauses 15.1 and 15.2 of Sub-Contract NAM/SC)* or determination of the main contractor's employment if earlier.

Clause 6.3B.2 is equivalent to clause 22B.2 of the main contract ('Failure of employer to insure – rights of contractor') with the addition of the words *'Except where the Employer is a local authority'* as an introductory statement to 6.3B.2. In Amendment 2 JCT 80 the equivalent clause 22B.2 is not used in the local authorities versions of JCT 80.

Clause 6.3C.2 is equivalent to clause 22C.2 of the main contract ('Works in or extension to existing structures – all risks insurance – employer to take out joint names policy') with the addition of the introductory statement mentioned in the previous paragraph, and the last sentence of this previous paragraph also applies to clause 22C.2.

There are two standard forms of sub-contract for use with IFC 84. The JCT produces a standard form of named sub-contract, NAM/SC, which must be used on every named sub-contract, as mentioned above. There is a standard form of tender which goes with it called NAM/T. The Building Employers Confederation has published a standard form of domestic sub-contract for use on such sub-contracts let under IFC 84, known as IN/SC. This sub-contract is almost identical to DOM/1. See Chapter 15.

16.3 JCT Prime Cost Contract 1992 Edition

This contract, which was first published in March 1967, provided for the remuneration of the contractor by repaying him his costs and in addition a fixed fee for his services. The contract is issued under the same associated bodies which issue the Standard Form of Building Contract. The Prime Cost form is used comparatively rarely as it is intended for circumstances where the work has to commence before its extent has been discovered, e.g. the repair of damage by fire. Thus the exact nature of the work is not known at the time the contract is executed. The 1992 edition for the Injury, Damage and Insurance section uses clause 6 as the IFC 84 form does, but the wording of the Standard Form is followed even more closely than that used in IFC 84. For example it was explained that there is no provision in IFC 84 for the employer to take possession of part of the works before practical completion unless advantage is taken of the additional clause mentioned. However, there is provision in the Prime Cost contract for partial possession as there is in the Standard Form.

As the wording of this Prime Cost Contract follows so closely that of the Standard Form it is possible to keep the commentary short and thus avoid repetition of the comments already made under the Standard Form. It is therefore suggested that reference is made to the table given when commenting on the IFC 84 form in order to ascertain the equivalent clause in the Standard Form and then refer to the commentary earlier in this book in the appropriate chapter dealing with that clause.

Another very slight difference from the Standard Form and IFC 84 appears by virtue of the nature of this Prime Cost contract. Provision is made for the employer to take out insurance which the contractor defaults in providing and to charge the contractor for this. However, the method of doing so is different in the Prime Cost contract as this charge has to be deducted from the Prime Cost. Similarly where the contractor's recovery for his outlays is limited to the insurance monies in the Standard Form and IFC 84, it is limited to the Prime Cost in this contract. There is a special version of the nominated sub-contract for use with the Prime Cost contract (NSC/PCC) which is exactly the same as NSC/C except for the references to the main contract which concern the Prime Cost contract instead of JCT 80.

Chapter 17

JCT Standard Form of Building Contract with Contractor's Design

17.1 Design and a work and materials contract at common law

As Chapter 1 dealing with legal liabilities did not deal with the legal aspects of design work, it is necessary to start with some basic legal facts.

Section 14 of the Sale of Goods Act 1979 (as amended) (subject to certain circumstances) imposes implied terms of 'satisfactory quality' (see Chapter 1) and that the goods are 'reasonably fit for the purpose required' in respect of goods supplied under contract, which terms are irrespective of negligence. By the Unfair Contract Terms Act 1977, section 6 imposes obligations in respect of consumer goods and states that the implied terms of sections 13 to 15 of the Sale of Goods Act 'cannot be excluded or restricted by reference to any contract term' as against a consumer, and can only be excluded against others if the exclusion is reasonable (see Appendix 5).

However, *construction work is not sale of goods* but a work and materials contract, although a contractor would be caught by section 7 of the 1977 Act (see also Appendix 5). Nevertheless, courts have imposed upon building work implied conditions similar to those mentioned above, as the following cases testify:

Lord Reid in *Young and Martin Ltd* v. *McManus Childs Ltd* (1969), quoted from the judgment of du Parcq J in *G.H. Myers & Co.* v. *Brent Cross Service Co* (1934):

'I think that the true view is that a person contracting to do work and supply materials warrants that the materials which he uses will be of good quality and reasonably fit for the purpose for which he is using them, unless the circumstances of the contract are such as to exclude any such warranty.'

Lord Reid said:

'The only addition I would make is that there are really two warranties, one as to quality and one as to reasonable fitness for the job.'

The case of *Hancock and Others* v. *V.W. Brazier (Annerley) Ltd* (1966) is of some assistance in attempting to solve the present problem. In that case

Lord Denning said that when a purchaser buys a dwelling from a builder who contracts to build it, there are three implied obligations on the builder:

> 'that the builder will do his work in a good and workmanlike manner;
> that he will supply good and proper materials;
> that the dwelling will be reasonably fit for human habitation.'

As regards the last obligation, Lord Scarman in *IBA* v. *EMI and BICC* (1981) did not limit it to dwellings as the article concerned was a mast. He said:

> 'In the absence of any term (express or to be implied) negativing the obligation, one who contracts to design an article for a purpose made known to him undertakes that the design is reasonably fit for the purpose.'

As far as contractors' duties are concerned this approach is consistent with the *Hancock* and *Young and Marten* cases.

The general result of these statements is that a design and build contractor's obligation is a strict obligation, irrespective of negligence, that the design will be fit for its purpose. In comparison, a designer, e.g. an architect or engineer (whose duty is only to use reasonable skill and care commensurate with the standard of an ordinary competent member of his profession at the time (see *Bolam* v. *Friern Hospital Management Committee* (1957)), has no such duty unless he chooses to accept the higher obligation just mentioned. See the first hearing of *Greaves & Co (Contractors) Ltd* v. *Baynham Meikle & Partners* (1975) explained by the Court of Appeal. Consequently, if the design-and-build contractor uses the reasonable skill and care of a designer, he will still be liable if the design is not fit for its purpose. Incidentally, there is no implied term of fitness by the main contractor, in his contract with the employer, for the design of a nominated sub-contractor, see *Norta Wallpapers (Ireland) Ltd* v. *John Sisk and Sons (Dublin) Ltd* (1978).

Regarding *Norta Wallpapers*, the House of Lords distinguished that decision in *IBA* v. *EMI and BICC* (1981) mentioned earlier (EMI had, but John Sisk had not, accepted liability for the sub-contractor's design). The question of EMI's responsibility for the design of the mast by BICC (in effect nominated sub-contractors) was considered by Lord Fraser who said:

> 'It is now well recognised that in a building contract for work and materials a term is normally implied that the main contractor will accept responsibility to his employer for materials provided by nominated sub-contractors. The reason for the presumption is the practical convenience of having a chain of contractual liability from the employer to the main contractor and from the main contractor to the sub-contractor – see *Young and Marten Ltd* v. *McManus Childs* ... Accordingly, the principle that was applied in *Young and Marten Ltd* in respect of materials, ought in my opinion to be applied her in respect of the complete structure, including its design. Although EMI had no specialist knowledge of mast design, and although IBA knew that and did not rely on their skill to any extent for the

design, I see nothing unreasonable in holding that EMI are responsible to IBA for the design seeing that they can in turn recover from BICC who did the actual designing.'

The best plan is for the main contract to make it clear that the sub-contractor is responsible for the design (see clause 58(3) of the ICE Conditions).

The Consumer Protection Act 1987 does not affect to any extent this area of the law as its importance lies in the extension of strict liability beyond the boundaries of the law of contract, and here we are involved in contract. Secondly, it was seen in Chapter 1 that the broad effect of the provisions of this Act seem to be that the supply by a builder of a building by virtue of the creation or disposal of an interest in land does not count as a supply for the purposes of the Act. This entitles the builder to apply the defence set out in section 4(1)(b) of the Act.

The exclusion of implied terms is caught by section 7 of the Unfair Contract Terms Act 1977 (see Appendix 5). This common law situation is now made a statutory one by the Supply of Goods and Services Act 1982.

17.2 Supply of Goods and Services Act 1982

This Act is divided into three parts:

- Part I, Supply of Goods;
- Part II, Supply of Services;
- Part III, Supplementary.

The definition of a contract for the supply of a service in this Act is a contract 'under which a person ("the supplier") agrees to carry out a service', and this applies whether or not goods are also transferred or to be transferred under the contract (sections 12(1) and 12(3)). Thus the contract is only defined in general terms. The effect is that the sections relating to supply of services will apply equally to all types of designers, e.g. architects, engineers, contractors, nominated sub-contractors (who often design) and suppliers carrying out design and build contracts.

Section 13 implies a term that the service will be carried out with reasonable skill and care. The Act does not prejudice any rule of law that imposes on the supplier a duty stricter than that of reasonable skill and care (section 16(3)(a)). Thus where a design and build contractor is operating under a contract which impliedly or expressly imposes on him an obligation that the design will be fit for its purpose, he will not be able to maintain that his stricter duty is reduced by the implied term of this Act.

The Act, under sections 11(1) and 16(1), further provides that a right, duty or liability which arises by reason of the provisions of the Act regarding the supply of goods or services can be negatived or varied by express agreement, or by the course of dealing between the parties, or by such usage as binds both parties to the contract. However such exclusions of liability are expressly made subject to the Unfair Contract Terms Act 1977 and in a design and build contract, for example, section 7 would probably operate (see Appendix 5).

Under sections 4(4) and 4(5) where the purpose of the goods is made known to the transferor, there will be an implied condition that the goods are reasonably fit for that purpose, whether or not it is a purpose for which the goods are commonly supplied. Where the transferee does not rely on the transferor's skill and judgment (or if it is unreasonable for him so to do) the implied term does not apply.

The conclusion is that this Act *inter alia* puts in statutory form the position given in *Young and Marten Ltd* v. *McManus Childs Ltd* and the other cases mentioned earlier. Consequently a design and build contractor is liable to produce materials of satisfactory quality and a product fit for its purpose, whereas the designer who does not supply goods is only liable if he fails to use reasonable skill and care. However these duties are subject to express exclusions, which in turn are subject to the Unfair Contract Terms Act 1977.

17.3 The Standard Form of Building Contract with Contractor's Design, 1981 Edition

This contract was published by the JCT together with two practice notes (CD/1A, which had a general description of the new form of contract including a page headed 'Insurance: Contractor's Design Liability', and CD/1B, which had detailed notes on the new form). This contract is used when the employer has stated in writing his requirements and the contractor has produced the design and tendered for the work. The pattern of this contract follows that of the main form of contract but the actual wording varies considerably. However, this considerable variation does *not* apply to the main liability (indemnity) responsibility for the works and complementary insurance clauses 20 to 22, the wording and numbering of which follow almost exactly the main JCT 80 (see later).

The design warranty clause 2.5 requires careful consideration. A noteworthy feature is that this contract does not include any requirement that the liability of the contractor under clause 2.5, concerning his design warranty, shall be insured.

Contractor's Design Warranty Clause 2.5

2.5.1 Insofar as the design of the Works is comprised in the Contractor's Proposals and in what the Contractor is to complete under clause 2 and in accordance with the Employer's Requirements and the Conditions (including any further design which the Contractor is to carry out as a result of a Change in the Employer's Requirements), the Contractor shall have in respect of any defect or insufficiency in such design the like liability to the Employer, whether under statute or otherwise, as would an architect or, as the case may be, other appropriate professional designer holding himself out as competent to take on work for such design who, acting independently under a separate contract with the Employer had supplied such design for or in connection with works to be carried out and completed by a building contractor not being the supplier of the design.

2.5.2 Where and to the extent that this Contract involves the Contractor in taking on work for or in connection with the provision of a dwelling or dwellings the reference in clause 2.5.1 to the Contractor's liability includes liability under the Defective Premises Act 1972 and where the application of section 2(1) of the Act is included in the Employer's Requirements the Contractor and the Employer respectively shall do all such things as are necessary for a document or documents to be duly issued for the purpose of that section and the scheme approved thereunder which is referred to in Appendix 1.

2.5.3 Where and to the extent that this Contract does not involve the Contractor in taking on work for or in connection with the provision of a dwelling or dwellings to which the Defective Premises Act 1972 applies, the Contractor's liability for loss of use, loss of profit or other consequential loss arising in respect of the liability of the Contractor referred to in clause 2.5.1 shall be limited to the amount, if any, named in Appendix 1: provided that such limitation of amount shall not apply to or be affected by any damages which under clause 24.2 the Contractor could be required to pay or allow at the rate stated in Appendix 1 as liquidated and ascertained damages in the event of failure to complete the construction of the Works by the Completion Date.

2.5.4 Any references to the design which the Contractor has prepared or shall prepare or issue for the Works shall include a reference to any design which the Contractor has caused or shall cause to be prepared or issued by others.

The effect of clause 2.5.1 is to make, or attempt to make, the contractor's liability in respect of design and same as that of the professional designer (e.g. the architect), that is, a liability only to use reasonable skill and care. Some doubts have been expressed as to whether this wording is sufficient to avoid the application of the implied terms mentioned earlier, viz. good workmanship, good materials, and fitness for the purpose required. Normally, express terms (as in the above clause) would override any implied terms if they clash. However, there are other tactics available to an employer who considers that this clause prevents him recovering damages from a contractor so far as that contractor's inadequate design is concerned. They are as follows.

Professional negligence

There is negligent design and defective design. The test of reasonable skill and care decides negligence. If the design has been defective but not negligent, as was the view of the Court of Appeal in *IBA* v. *EMI and BICC*, mentioned earlier, then it is necessary to decide whether the design responsibility extends to ensuring that the design was reasonably fit for the purpose for which it was required. The Court of Appeal did take this view of its extent and Lord Scarman argued vigorously in the House of Lords that this was correct in accordance with his statement given earlier. The other judges did not express a final view on this matter although Viscount Dilhorne commented that he would be surprised if the duty did not extend this far. It therefore seems that in the case of a design and build contract the contractor has to meet this extended duty in order to avoid professional negligence.

As it is easier to prove a breach of this extended, strict design duty than to prove negligent design it seems that the former will be used or become part of the latter in these circumstances. However, to refer again to Lord Scarman's statement, the introductory words and those in brackets (see the beginning of this chapter) indicate a proviso, and in that case of clause 2.5.1 the important question is whether it contains an express term negativing the obligation of fitness for purpose. The main area of controversy seems to concern the applicability of Lord Scarman's *dicta* to 'an architect or other appropriate designer', to use the words of clause 2.5.1. Lord Denning in *Greaves & Co (Contractors) Ltd* v. *Bayham Meikle & Partners*, mentioned earlier, said:

> 'What then is the position when an architect or an engineer is employed to design a house or a bridge? Is he under an implied warranty, that if the work is carried out to his design, it will be reasonably fit for the purpose? Or is he only under a duty to use reasonable care or skill? This question may require to be answered some day as a matter of law. But in the present case I do not think we have to answer it.'

Defective Premises Act 1972

See Chapter 1. Section 1 of this Act applying to dwellings imposes the implied terms mentioned earlier, viz. work must be done in a professional manner with proper materials so that the dwelling will be fit for habitation, and there can be no contracting out of these terms. So, even if clause 2.5.1 achieves its object of avoiding the implied terms, the Act will impose them so far as dwellings are concerned. However, note the exception mentioned in Chapter 1 of an approved scheme although there is no such scheme at the time of writing. This explains the effect of clause 2.5.2, which makes provision for the 1972 Act, and the intention of the reference to section 2(1) of this Act is to provide in Appendix 1 of the JCT Design and Build contract for an approved scheme to apply if required.

Misrepresentation and negligent misstatement

Sometimes the term 'negligent misrepresentation' is interchanged with 'negligent misstatement'. However misrepresentation is a contractual matter whereas negligent misstatement arises in tort. Thus misrepresentation can only apply between the parties to the contract and negligent misstatement is not so limited. However, whereas the former is a false statement of fact made at the time the contract is entered into and is a reason why the recipient makes the contract, the latter is governed by the case of *Hedley Byrne & Co Ltd* v. *Heller & Partners Ltd* (1963) and involves a special relationship between the parties (explained in *Esso Petroleum Co Ltd* v. *Mardon* (1976)) whereby a duty exists not to make a negligent statement.

The result of misrepresentation is that the victim of the false statement can insist on performance of the contract or rescind the contract. The Misrepresentation Act 1967 allows the innocent party to sue in damages for any loss suffered unless the other party can show he had reasonable grounds for believing, and did believe when making the contract, that the statement was true. Consequently, the employer could, if he satisfied the above requirements (including a representation that the contractor's design would be fit for the employer's requirements, as they are called, in the contract), bring an action for damages.

Clearly, under the *Hedley Byrne* decision, the designer could be liable for negligent statements about the design.

The Unfair Contract Terms Act 1977 (see Appendix 5)

As Clause 2.5.1 restricts the contractor's liability, the question is whether the JCT Design and Build contract under section 3 of the Act 'is the other's written standard terms of business', and there is a doubt about this, particularly where the employer belongs to one of the representative bodies on the JCT. In the case of a private employer on a one-off contract, it might be considered that the contract is the contractor's standard terms of business. In this event the contractor must satisfy the reasonableness test. Also the limit of liability mentioned in clause 2.5.3 and set out in Appendix 1 of the contract would have to be reasonable. Between two commercial organisations it is likely that these terms would be considered reasonable, particularly when they are agreed by the representative bodies of both parties. See the suggestion of this in *Walker* v. *Boyle* (1982).

Clauses 2.5.3 and 2.5.4

Clause 2.5.2 has been dealt with under the 'Defective Premises Act' heading mentioned earlier.

Clause 2.5.3 is the alternative to clause 2.5.2 where the work does not consist of dwellings to which the 1972 Act applies and a limitation can be placed on the contractor's liability under Appendix 1. It is not contrary to this Act to limit liability under sections 3 and 4 of the Act. Clearly, in view of the extent of the contractor's liability (including loss of use, loss of profit and other consequential loss), the limit should be a high one.

Clause 2.5.4 incorporates into the contractor's design any reference to any design prepared or issued by others. Thus, *vis-à-vis* the employer, the contractor is liable for all such designs but may be able to exercise recovery in certain circumstances.

17.4 Liability, responsibility for the works and complementary insurance clauses (20 to 22) of the Design and Build contract

The JCT contract 'with contractor's design', issued in 1981, is referred to as JCT 81. The revised insurance and liability provisions are set out in 'Amendment 1 issued November 1986' for JCT 81, and these largely follow those in Amendment 2 issued November 1986 for JCT 80. The following differences now apply (bearing in mind that the clause numbers are almost exactly the same as in the main JCT contract).

(1) The word 'Employer' replaces the word 'Architect' as the Design and Build contract makes no provision for an architect or quantity surveyor by its very nature.

(2) The employer's requirements and Appendix 1 may make provision for a clause 21.2.1 insurance, and, although no provision is made for the premium for this insurance to be added to the JCT contract sum as in the main JCT form clause 21.2.3, this is unnecessary in JCT 81 as the cost of this insurance is already included in the contract sum.

(3) The footnote in the 1986 version of JCT 80 (concerning the non-standardisation of 'all risks' insurance) is reproduced as footnote [h.1] in the JCT 81 amendment, with the following sentence added:

> In particular for contracts under this Form [i.e. JCT 81] cover may only be given subject to design in the definition in clause 22.2, paragraph 2 of All Risks Insurance being an absolute exclusion.

Probably the contractor's all risks policy will have an exclusion at least limiting it to a defect in design in the work or materials themselves. However, this is a material damage policy and if the professional indemnity policy (a liability policy) is taken out to cover the warranty on design given by the contractor under JCT 81 clause 2.5 the overlap is small (limited to the works which are not defective in design). In any event, the contractor's design liability is not intended to exceed that of the professional designer (the architect if there was one) (see clause 2.5.1).

(4) As there are no nominated sub-contractors under JCT 81, clause 22.3 concerning loss or damage by the specified perils, only gives to sub-contractors the benefit of the joint names policies under clause 22A, or 22B or 22C.2 for the works, and not for the benefit of the joint names policy for the loss or damage to the existing structures and their contents under clause 22C.1. It should be appreciated that all sub-contractors are domestic under JCT 81.

Insurance requirements of the Design and Build contract

Clauses 20 to 22 require the same insurance policies mentioned in the previous chapters of this book. It is the insurance suitable for clause 2.5 that requires some explanation. In the first place one cannot do better than quote from page 12 of Practice Note CD/1A:

'30 The Form of Contract does not include any requirement that the liability of the Contractor under clause 2.5 shall be insured.

31 If the Employers wish to consider such an insurance obligation it would be necessary to make appropriate provision in the "Employer's Requirements" which in turn would have to be matched by proposals in respect of insurance in the "Contractor's Proposals".

32 Great care has to be exercised that any insurance requirement corresponds with what is possible of negotiation in the insurance market. Insurance for the type of liability imposed upon the Contractor in clause 2.5 would normally be provided by what is known as a "Professional Indemnity" policy. It is important for both Employers and Contractors to note that such a policy will in no circumstances provide indemnity for simple inadequacy or ineffectiveness of design, *merely the Contractor's legal liability for neglect omission or error in the exercise of the professional function equivalent to that which would normally fall to an Architect or similar consultant.* In short, broadly, the Contractor's negligence.

33 For these reasons a demand for insurance equating to a total guarantee of the design in all circumstances is not practicable nor (leaving aside the general unwillingness of the insurance market to provide it) is such a demand in line with the liabilities accepted by the Contractor under clause 2.5. It follows that neither will a Professional Indemnity policy provide any indemnity whatsoever in respect of defective materials or workmanship.

34 If they do decide to seek insurance cover, Employers should also give consideration to the mechanics by which this is provided. It is anticipated that they will find themselves faced with one of two situations:

(a) the Contractor will produce an annually renewable Professional Indemnity policy which he has already in being to provide protection in respect of his design activities. Such a policy would normally be so drafted as to include all work undertaken during the year of insurance and, provided the Contractor renews it annually, should be fully retroactive in effect, i.e. provided it is kept in force, it will not matter when the design error was committed or when a claim is made; or

(b) the Contractor will offer a "one-off" Professional Indemnity policy specifically for the contract under consideration.

35 It is important to note, especially in view of recent legal decisions, that causes of action in tort against the Contractor may accrue at a

considerable time after completion of the building works, and the limitation period runs thereafter. A policy of the type envisaged in paragraph 34(b) above will normally provide indemnity *solely in respect of claims made or incidents notified during the period of construction plus defects.*

36 Although in some circumstances it may be possible to extend the liability period, the insurance market will not grant cover beyond a total period of six years at the most. As stated earlier, an annually renewable policy if kept in force, should provide continuing cover subject otherwise to its terms and conditions.

37 On the other hand, where the Contractor infrequently engages in design work it is scarcely practicable to expect him to maintain permanently in force an annually renewable policy covering perhaps only a single contract. It may be possible in such circumstances for Employer and Contractor to agree a point in time when the Contractor's liability ceases, but this is a matter which a Practice Note must leave to the parties concerned.

38 Two points to be watched carefully are: that the cover operates fully during the construction period as well as thereafter; and that the Contractor is indemnified for liability arising from the professional error of any party to whom he may decide to sub-let design work.

39 The need for skilled professional advice concerning Professional Indemnity insurance is clear. Particularly is this so in that the Professional Indemnity insurance market is relatively limited in size, the amounts potentially at risk may be very large indeed, and there are important differences in the cover offered by various sectors of the insurance market.'

Comments arising from Practice Note CD/1A, page 12

Whether the contractor elects for an annually renewable or a 'one-off' policy, the premium calculation is affected by various factors, e.g.

- the amount insured;
- professional qualifications and experience of the contractor's in-house design team;
- the insurance cover carried by any firm to whom design activity is let and the amount thereof;
- the percentage of work value designed as compared with total project value (or annual turnover if a renewable policy);
- and most important, the type of work to be designed.

A proposal form will almost certainly be required, and it is of the utmost importance that it is fully and accurately answered if the danger of a policy claim being rejected is to be avoided.

In the case of annually renewable policies, such a proposal form is likely to be required each year. This tends to be a complicated document that is not easy to answer comprehensively, especially in regard to estimating the degree of design likely to be executed under different headings in the forthcoming year. The guidance of the contractor's insurance advisers is strongly recommended.

It is essential that claims upon the contractor be notified immediately to the insurers, to avoid a plea that their interests have been prejudiced. In some cases this may be difficult as considerable time may elapse before there is any clearly definable implication that the defect is due to defective design.

The lesson to be learnt from all this is that the contractor should sensibly err on the side of caution and advise his insurers immediately of anything which even remotely may involve an allegation of defective design. Admittedly the drawback to such an approach is that when the policy is due for renewal there will be these queries upon the contractor's record, which could have an effect upon the renewal premium charged. Nevertheless, it remains the lesser of two evils.

A necessary corollary to the above is that the contractor should not go ahead and make good a defect without his insurers' consent, since this may result in important evidence being destroyed or covered up. Apart from provoking an alleged prejudice to their position, such action may be construed by the insurers as an admission of liability and therefore a breach of a policy condition.

It is of vital importance that a professional indemnity policy fully and correctly describes all the contractor's professional activities, since it will be only in respect of these that he is indemnified. In the context of design and build work, it will often be necessary to ensure that such description includes not merely 'designers' but 'project managers and/or supervisors' if this is in effect what the contractor's function is, as it often will be.

The more important general points concerning the cover provided by the professional indemnity policy are given in Section 17.5 and the application of that policy to a design and build department of a contractor is considered in Section 17.6.

17.5 Professional indemnity insurance

(1) A professional indemnity policy protects the insured against his legal liability to pay damages to third parties who have sustained some injury, loss or damage due to the negligence of the insured or his staff in the professional conduct of the business.

(2) Generally speaking, liability is based on breach of professional duty which is the failure to exercise an ordinary degree of care and skill in the profession concerned. There is only a limited insurance market available and no standard policy form. A typical operative clause reads:

'To indemnify the insured against any claim for damages for breach of professional duty which may be made against him during the period of insurance due to any negligent act, error or omission whenever or wherever the same was or was alleged to have been committed by the insured or his predecessors in business or any employee of the insured or his predecessors in business in the conduct of the business.'

(3) The claim has to be made against the insured *during the period of insurance,* irrespective of when the act, error or omission took place.

(4) The amount up to which insurers will pay claims is stated as a limit of indemnity on the policy. Usually the limit applies in the aggregate, thus applying to all claims in any one policy year. There is a substantial excess.

(5) In the usual basic form of policy cover the policy exceptions are few. Claims arising from libel or slander; dishonest, fraudulent, criminal or malicious acts or omissions of the insured or his employees; and claims in respect of which the insured is entitled to indemnity under another policy, are excluded. However, the first two exclusions may be removed on payment of an additional premium, but not in respect of the dishonest, fraudulent, criminal or malicious act or omission of the insured himself, only of his employees.

(6) Three policy conditions call for comment:

- The condition concerning the control and negotiation of claims being limited to the insurers contains a QC clause. This states that the insured shall not be required to contest any legal proceedings unless a Queen's Counsel (mutually agreed upon by the insurers and the insured) shall advise that such proceedings should be contested.
- Notification of claim. The insurers should be brought in as soon as there is any suggestion of a claim.
- Notification of occurrence. If the insured becomes aware of an occurrence which may give rise to a claim against him, he must give immediate notice of the incident. If this is done, any such claim which arises from that occurrence shall be deemed to have been made during the subsistence of the policy.

(7) The policy can be extended to cover libel and slander, the dishonesty of employees, and the liability of incoming and outgoing partners.

(8) A case which shows there is potentially a wider liability for architects or engineers, when this policy would come into consideration, is *Crown Estate Commissioners* v. *John Mowlem & Co Ltd* (1995). Clause 30.9.1 of JCT 80 provides that the Final Certificate shall be conclusive evidence that, where the quality of materials or standard of workmanship are to be to the reasonable satisfaction of the architect, the same are to such satisfaction. One question for the court in this case was whether the Final Certificate only applies to standards and quality expressly to be to the architect's satisfaction (the narrow construction) or on the quality of all materials and the standard of all

workmanship (the wider construction). The wider construction is correct. This wider construction would include conformance with the criteria stated in the contract documents, e.g. British Standard Specifications, and implied terms that materials must be of satisfactory quality and fit for their purpose. This decision follows *Colbart* v. *Kumar* (1992) which concerned the JCT Intermediate Form. Thus, as in this case, a building owner who belatedly discovers his building is defective will not have a remedy against the contractor if the Final Certificate under the JCT form has been issued. His only remedy will be against the architect or engineer, who could be accused of failing to identify defective work and by issuing a Final Certificate has prejudiced the building owner.

(9) Under the next heading the application of this policy (among others) to the design and build department of a contractor is considered.

17.6 Types of insurance which might be relevant to the design and build risk

Contractors' all risks

Some contractors might be under the impression that their CAR policy will cover them in respect of design work. While it has been explained in Chapter 7 that a limited type of design cover is given by some insurers, namely the damage due to the defective design, this is limited to work which is not itself defectively designed (although JCT clause 22.2 also excludes work supported by defective design). Even if, for example, an internationally known contractor with considerable influence can obtain full design cover under the CAR policy, the indemnity will lapse at the end of the construction period plus the defects liability or maintenance period, and design defects have a tendency to appear a long time after completion of the work. Furthermore consequential loss cover, unless specially arranged, is excluded from the CAR policy. So far as this policy is concerned, insurers feel that to cover design risks removes the necessity for the contractor to maintain standards of design, quality and work performance. They fear that risky, novel and untried features and experiments will be carried out, the failure of which is probable, and for which they should not be expected to pay. Apart from this, they would be overlapping the cover of the professional indemnity policy. Incidentally the ABI have suggested five wordings for this exclusion which vary according to the extent to which the insurer is prepared to cover the design risk.

Public liability

These policies (even with a products liability extension) will exclude damage to the product or thing sold or supplied, and design risks are often specifically excluded, except where no fee is charged. Damage to the contract works are excluded by various different wordings but the effect is largely the same. However, while the cost of repairing defective work is not covered, the consequential damage (other than to the works) resulting from the defective work is covered. Clearly the public liability policy insurer does not intend to cover the risk of professional negligence, advice or design, but if the policy does not contain such an exclusion the other exceptions mentioned above will put some limitation on design risk cover.

Moreover, the restriction of the operative clause to injury to persons and damage to property will exclude the pure financial or economic loss claim by a third party which does not arise out of physical injury or damage to property. Thus if an owner of land developing a housing estate leaves the contractor's design and build department to prepare the plans, which contain a defect creating such a delay that the whole project has to be abandoned and the land sold, the financial loss sustained by the owner would be the responsibility of the contractor. However, his public liability policy would not apply as the claim does not involve bodily injury or damage to property.

As was explained in Chapter 6 the operative clauses of some contractors' public liability policies will cover accidental obstruction, trespass or nuisance, which would cover a pure economic loss claim arising from these risks. On the other hand the cost of making good defective or faulty work or materials would probably be excluded as will damage to any structure supplied or erected by the contractor.

Professional indemnity

While this policy (from among those most available to the contractor) comes nearer than any other to providing cover for the design risk, the following aspects must be noted:

(1) There is often some form of contractual liability exclusion as the intention is to cover only breach of professional duty. In fact the giving of any express or even implied warranty which increases the insured's liability is often a specific exception, emphasising that only professional negligence is covered and not a pure breach of contract (see the explanation of *Greaves & Co (Contractors) Ltd* v. *Baynham Meikle & Partners* (1975) given in Section 17.1). The result is that this policy does not cover the implied conditions mentioned at the beginning of this chapter that the completed work will be of good quality and materials and suitable for its intended purpose. This goes beyond professional negligence and calls for a products guarantee policy, but see later the difficulties in obtaining such cover.

(2) Professional negligence policies are usually 'tailor made' for the design department of a contractor by an insurance broker who specialises in catering for the construction industry. Such a design department would include unqualified as well as qualified persons, although basically this policy is for qualified people. However the policy may stipulate that the

design, specification and/or management of the work described in the schedule of the policy is to be under the direction of qualified architects, engineers, surveyors and/or project managers.

(3) These policies may contain the following terms which are often negotiable:

- A large excess (or deductible, as it is sometimes called) in the order of £10,000, or more.
- A comparatively low limit of indemnity for each and every claim and in all.
- An exclusion of claims arising from the injury, disease, sickness or death of any person (or just employees) which would exclude the type of claim which arose in the case of *Clay* v. *Crump* (1963).

(4) If this policy is limited to cover the design risk and the materials used were faulty because they were the wrong materials to specify, this would be a design fault and covered by the policy. But if they were defective in themselves (not up to standard), this is not a design fault and not covered by the policy. However, the implied terms of good quality and materials and fitness for the purpose would impose a strict liability in both cases (see the *Hancock* case mentioned earlier).

(5) As this policy is on a claims-made basis the employer will have to check the policy period and whether it has retroactive cover. This arises where protection is required for work carried out before the policy cover commenced. For example, cover may be taken out on 1 January 1995, which usually means that cover is for claims made during the period of insurance, in respect of work done also in that period. However, if protection is required for claims arising in the period of insurance from work done in earlier years then a retroactive date has to be arranged at an additional premium. If run-off liability cover is required usually the insurance cover is limited to a period of six years maximum, whereas legal liability under the Latent Damage Act 1986 provides a long-stop period of 15 years from the breach of duty (if this expires before the basic periods of six years from the occurrence of the damage or three years from its discoverability, whichever is the later) when the damage is statute barred.

(6) An attempt may be made by means of a collateral warranty, to impose on the contractor, or even a sub-contractor involved in design work, a requirement to take out professional indemnity insurance for up to fifteen years after practical completion of the works. It would be very unwise of the contractor or sub-contractor to enter into such a legal obligation in view of what has been said about the availability of such insurance cover. A possible amendment to this type of warranty is that the insurance should be continued only as long as it remains available at a reasonable rate of premium.

Products guarantee

This is the type of cover that is required to meet a design risk. It protects the insured in respect of his legal liability to replace or repair the defective or faulty product giving rise to the claim, and against its failure to fulfil its intended function, including legal liability for financial loss sustained as a consequence. An insurance broker catering for the construction industry should be consulted, but it is nevertheless very difficult for the industry to obtain this cover, particularly for the construction of dwellings. Furthermore, even if it is obtainable, it is very expensive and there is usually an excess in the shape of a percentage of the loss being retained by the insured as his own liability.

Decennial insurance or building guarantee insurance

This is a type of cover which gives some protection to the building owner against major structural damage, but only one or two insurers offer this cover, and generally speaking not in respect of dwellings. Unfortunately this insurance is not for the contractor, but the owner can, by paying an additional premium, waive his subrogation rights against the contractor which gives the latter some form of protection. However there is really no reason why the owner should do this.

National House-building Councils' scheme

This scheme protects the purchaser of a new home against defects in its design or construction but there is no protection for the contractor so there is no point in going into any detail.

Performance bond

It is sometimes thought that a performance bond might help. See Chapter 19. However, the contract does not ask for a bond and moreover the following points are against such protection:

- The bond amount is usually 10% of the contract price and this could be quite inadequate.
- This bond is for the due performance of the works and it is not a guarantee to last after completion.

Most bonds used by insurers provide for their release either at the date of expiration of the period of maintenance, or on the issue of a certificate of completion of the works by the architect or of substantial completion of the works by the engineer, whichever shall be the last to occur. If there is no release clause, it is possible that some cover is available as long as the bond remains in force. However, Rowlatt on *Principal and Surety*, 4th edition (1982) has the following to say on the subject:

'The usual rule is that when the principal debtor has complete his performance the surety is discharged. In building terms the performance may be "complete" but the building contract may require the architect to certify certain stages of completion by means of certificates, e.g. as in the JCT form. Where a building contract, unlike the JCT form, does not specify any stages of certificate completion, actual completion may be taken as complete performance by the contractor of the principal obligation so as to discharge the surety (*Lewis* v. *Hoare* (1881)). Where the building contract specifies certain stages of certified completion but the guarantee is silent as to which stage is meant, it is submitted that the Final Certificate should be taken as the stage of complete performance for the purpose of the guarantee.'

Conclusion

(1) The insurance market will provide cover for the professional negligence risk of a design and build contract if the right section of the market is approached. Most professional indemnity insurances are arranged for contractors through Lloyds and with a few companies.

(2) Generally speaking, the insurance market does not cater for the cover required for breach of the implied terms of good workmanship, materials and fitness for purpose except in a limited way. The insurance of dwellings in this respect is particularly difficult if not impossible to obtain.

(3) In *Wimpey Construction UK Ltd* v. *Poole* (1984), the plaintiff construction company claimed that it had been negligent in respect of its design and was therefore entitled to be indemnified by the defendant insurer under a professional negligence policy. The judge followed the test set out in the *Bolam* case (mentioned earlier in this chapter) but he also held that *if* a professional man in fact had a higher degree of knowledge or awareness than the Bolam test required and acted in a way which, in the light of that knowledge, he ought reasonably to have seen would cause damage, he would be liable in negligence even though the ordinary skilled man would not have had that knowledge.

In the same case it was held that the operative clause of the policy expressed to cover 'omission, error or negligent act', covered an omission or error without negligence even though an ancillary clause referred to 'negligent act error or omission'. It was also stated that the cases relied on in *MacGillivray and Parkinson on Insurance Law*, 7th edition (1981), at paragraph 2024 did not support its assertion that the words 'negligent act error or omission' were apt to cover only negligence. While this is a step towards covering a breach of the implied conditions mentioned earlier (subject to no specific policy exclusion existing which concerned these conditions) it was also stated that not every loss caused by an omission or error was recoverable under the policy. It must be one which in principle could create liability (see *Haseldine* v. *Hosken* (1933)) and must not be deliberate.

There were two other aspects of this *Wimpey* case. It was also held that:

- In determining whether a structure had been negligently designed it was necessary to apply the professional standards of the time at which it was designed, not only later standards.
- The basic test of the same insured, the same property, and the same risk must exist before contribution between insurers applied to 'double insurance' (see *Petrofina Ltd* v. *Magnaload Ltd* (1983) in Chapter 2).

17.7 JCT Contractors' Designed Portion Supplement 1981

This Supplement prepared by the JCT for use with the Standard Form of Building Contract with Quantities (private and local authorities 1980 editions) provides for the contractor to be responsible for the design and construction of *part of the works.*

The Supplement consists of Articles of Agreement, additional conditions, amendments to JCT 80 conditions and a Supplementary Appendix. As the individual amendments and new conditions in the main followed the design and build form considered in this chapter, it is not proposed to comment any further. Most of the commentary in this chapter also applies to this Supplement. However, attention is drawn to Practice Note CD/2 concerning this Supplement.

17.8 Domestic sub-contractors and the DOM/2 sub-contract

The DOM/2 form is the sub-contract for use where the main contract is the JCT Contract with Contractor's Design 1981. The main distinction between the DOM/1 and the DOM/2 forms is the addition of a clause concerning design. Thus clause 5.3.1 of DOM/2 uses the equivalent wording used in clause 2.5.1 JCT 81, requiring the sub-contractor to take the same amount of skill and care required of the usual professional designer, i.e. the architect or engineer. The purpose is to remove any obligation of fitness for purpose which might otherwise be imposed on the sub-contractor (see earlier in this chapter).

Chapter 18

JCT Management Contract 1987 edition and Measured Term Contract 1989 edition

18.1 The JCT Management Contract 1987 Edition

The purpose of the management contractor is to supervise for the employer the works contractors, who perform the contract. Consequently there is a standard form of management contract between the employer and the management contractor. Practice Note MC/1 recommends the following as suitable conditions for the management contract, namely where:

- the employer wishes the design to be carried out by an independent architect and design team;
- there is a need for early completion;
- the project is fairly large;
- the project requirements are complex;
- although the employer requires early completion, he wants the maximum possible competition in respect of the price for the building works.

The word 'project' is used for the building work required by the employer and the term 'works' is used for the building work and services installations to be completed by the works contractors.

The architect and design team carry out the design of the project. While the management contractor advises on the practical aspects of design, he is not responsible for the design. Also the management contractor is prima facie responsible for the defaults of the works contractors, but is relieved of that liability under a complex set of relief provisions. The result is that the elements of the management contract are that:

- it is a low risk for the management contractor;
- it is suitable for large and complicated projects where speedy completion is required; and
- it is a prime cost more than a lump sum contract.

There is a standard form of works contract between the management contractor and the various works contractors under which the works for the project are carried out. This is in two parts namely Works Contract/1 and Works Contract/2. Works Contract/1 is the invitation to tender from the management contractor, a completed tender by the works contractor, and the recitals and articles of agreement. Works Contract/2 is the works contract conditions, which are incorporated into the articles.

There is also a Works Contract/3 which is an agreement between the employer and the works contractors containing certain rights and obligations which apply directly between those parties, including the creation of a direct liability to the employer in respect of design.

The Insurance Terms of Management Contract JCT 87

It is interesting to note that the general pattern of the liability, responsibility for the works and the complementary insurance clauses in JCT contracts is to deal with the liability for third party injury or damage to property before the responsibility for the works, and this is so even if one clause is handling both aspects as in the management contract. However, while clause 6 of JCT 87 deals with the insurance and indemnity provisions, the first sub-clauses handle the responsibility for the works, and the third party aspect is dealt with in the later sub-clauses.

Clause 6.1 states that clause 6.4 (all risks insurance of the project – management contractor to take out and maintain joint names policy) applies whether or not clause 6.5 (existing structures – insurance) applies. See below.

Clause 6.2 follows exactly clause 22.2 of JCT 80 as amended in 1986, i.e. it gives the wording of the operative clause and exceptions of the all risks insurance required in the following sub-clauses concerning responsibility for the works, plus the usual definition of 'Site Materials'.

Clause 6.3 provides for either a joint names policy to be taken out by the management contractor in accordance with clauses 6.4.1.1 or 6.4.3.1 (see later) or to be taken out by the employer where clause 6.5.2 is applicable (see later). In each case the policies are taken out in the joint names of the employer and the management contractor. This clause then follows clause 22.3.1 of JCT 80 as amended in 1986 reading 'Works Contractor' for 'Nominated Sub-Contractor' and referring to the Works Contract Conditions instead of NSC/C. Thus the works contractor is covered by this CAR policy but only for specified perils, as the sub-contractor is in JCT 80 as amended.

Clauses 6.4 and 6.5 read as follows:

> **[g] All Risks Insurance of the Project – Management Contractor to take out and maintain Joint Names Policy.**

> 6.4.1.1 The Management Contractor shall, prior to the commencement of any work on site for the Project, take out a Joint Names Policy for All Risks Insurance cover no less than that defined in clause 6.2 [h][k.1] (or for such other definition of cover as the Employer may instruct) for the full reinstatement value of the Project (plus the percentage, if any, to cover professional fees stated in the Appendix) and shall (subject to clause 2.8.3) maintain such Joint Names Policy up to and including the date of issue of the certificate of Practical Completion or, where the Project does not comprise alterations or extensions to existing structures, up to and including the date of determination of the employment of the Management Contractor (whether or not

the validity of that determination is contested) under clauses 7.1 to 7.13 or, where the Project comprises alterations of or extensions to existing structures, under clause 6.4.8 or clauses 7.1 to 7.13, whichever is the earlier.

.2 The Management Contractor shall, before taking out the Joint Names Policy, notify the Architect who shall thereupon notify the Employer of the amount of any excess (uninsured amounts) in respect of each insurance risk stated in the Policy. Subject to any alteration to such amounts of excess which the Employer may require and the insurers agree, the amounts of any excess in respect of each insurance risk insured under the Joint Names Policy shall be set out in the Appendix Part 2.

.2 The Management Contractor shall send to the Architect/the Contract Administrator for deposit with the Employer the Joint Names Policy referred to in clause 6.4.1.1 and the premium receipt therefor and also any relevant endorsement or endorsements thereof as may be required to comply with the obligation to maintain that Policy set out in clause 6.4.1.1 and the premium receipts therefor.

.3.1 If the Management Contractor independently of his obligations under this Contract maintains a policy of insurance which provides (inter alia) All Risks Insurance for cover no less than that defined in clause 6.2 [h][k.1] (or for such other definition of cover as the Employer may instruct) for the full reinstatement value of the Project (plus the percentage, if any, to cover professional fees stated in the Appendix) and the Employer has given to the Management Contractor his written acceptance of the amount of any excess in respect of each insurance risk stated in the policy (which amounts shall be set out in the Appendix Part 2) then the maintenance by the Management Contractor of such policy shall, if the policy is a Joint Names Policy in respect of the aforesaid Project, be a discharge of the Management Contractor's obligation to take out and maintain a Joint Names Policy under clause 6.4.1.1.

.2 If and so long as the Contractor is able to send to the Architect/the Contract Administrator for inspection by the Employer as and when he is reasonably required to do so by the Employer documentary evidence that such a policy is being maintained then the Management Contractor shall be discharged from his obligation under clause 6.4.2 to deposit the policy and the premium receipt with the Employer but on any occasion the Employer may (but not unreasonably or vexatiously) require to have sent to the Architect/the Contract Administrator for inspection by the Employer the policy to which clause 6.4.3.1 refers and the premium receipts therefor.

.3 The annual renewal date, as supplied by the Management Contractor, of the insurance referred to in clause 6.4.3.1 is stated in the Appendix.

.4 If any loss or damage affecting work executed or any part thereof or any Site Materials is occasioned by any one or more of the risks covered by the Joint Names Policy referred to in clause 6.4.1.1 or clause 6.4.3.1 then, upon discovering the said loss or damage, the Management Contractor shall forthwith give notice in writing both to the Architect/the Contract Administrator and to the Employer of the extent nature and location thereof; and the provisions of clause 6.4.5 to clause 6.4.9 shall apply.

.5 The occurrence of such loss or damage referred to in clause 6.4.4 shall be disregarded in computing any amounts payable to the Management Contractor, whether or not in respect of work executed by a Works Contractor, under or by virtue of this Contract.

.6 After any inspection required by the insurers in respect of a claim under the Joint Names Policy referred to in clause 6.4.1.1 or clause 6.4.3.1 has been completed the Management Contractor with due diligence, shall subject to clause 6.4.8 where applicable, secure the restoration of work damaged, the replacement or repair of any Site Materials which have been lost or damaged, the removal and disposal of any debris and proceed with securing the carrying out and completion of the Project.

.7 The Management Contractor, for himself and for all Works Contractors who are, pursuant to clause 6.3, recognised as an insured under the Joint Names Policy referred to in clause 6.4.1.1 or clause 6.4.3.1, shall authorise the insurers to pay all monies from such insurance in respect of the loss or damage referred to in clause 6.4.4 to the Employer.

.8 Clause 6.4.8 applies only where the Project comprises alterations of or extensions to existing structures.

.1 If it is just and equitable so to do the employment of the Management Contractor under this Contract may, within 28 days of the occurrence of the loss or damage referred to in clause 6.4.4, be determined at the option of either party by notice by registered post or recorded delivery from either party to the other. Within 7 days by [sic.] receiving such a notice (but not thereafter) either party may give to the other a written request to concur in the appointment of an Arbitrator under section 9 in order that it may be determined whether such determination will be just and equitable;

.2 upon the giving or receiving by the Employer of such a notice of determination or, where a reference to arbitration is made as aforesaid, upon the Arbitrator upholding the notice of determination, the provisions of clause 7.6.2 except clause 7.6.2.5 shall apply.

.9.1 Where the restoration, replacement or repair of the loss or damage and (when required) the removal and disposal of debris is carried out by a Works Contractor or Works Contractors already engaged upon the Project such restoration replacement or repair and, when required, the removal and disposal of debris shall be treated as it they were the subject of a Works Contract Variation required by an instruction under clause 3.4.

.2 Where clause 6.4.9.1 is not applicable the Management Contractor shall secure the restoration, replacement or repair of the loss or damage and, when required, the removal and disposal of debris, by a Works Contractor who shall be appointed in accordance with an Instruction under clause 8.1 and treated in all respects as a Works Contractor.

Specified Perils – insurance of existing structures and contents – Employer to take out and maintain Joint Names Policy

6.5.1 Clauses 6.5.2 and 6.5.3 apply only where the Project comprises alterations of or extensions to existing structures.

.2 The Employer shall, prior to the commencement of any work on site for the Project, take out a Joint Names Policy in respect of the existing structures (which shall include from the relevant date any relevant part to which clause 2.8 refers) together with the contents thereof owned by the Employer or for which he is responsible, for the full cost of reinstatement, repair or replacement of loss or damage due to one or more of the Specified Perils [k.2] and maintain such insurance up to and including the date of issue of the certificate of Practical Completion or up to and including the date of determination of the employment of the Management Contractor under clause 6.4.8 or clauses 7.1 to 7.4 or clauses 7.5 and 7.6 or clauses 7.7 to 7.9 or clauses 7.10 to 7.13 (whether or not the validity of that determination is contested) whichever is the earlier. The Management Contractor, for himself and for all Works Contractors who are, pursuant to clause 6.3, recognised as an insured under the Joint Names Policy referred to in clause 6.5.2 shall authorise the insurers to pay all monies from such insurance in respect of loss or damage to the employer.

.3 The Employer shall, as and when reasonably required to do so by the Management Contractor, produce documentary evidence and receipts showing that the Joint Names Policy required under clause 6.5, has been taken out and is being maintained. If the Employer defaults in taking out or in maintaining the Joint Names Policy required under clause 6.5.2 the Management Contractor may himself take out and maintain a Joint Names Policy against any risk in respect of which the default shall have occurred and for that purpose shall have such right of entry and inspection as may be required to make a survey and inventory of the existing structures and the relevant contents.

Clause 6.4 thus provides that the management contractor shall take out a joint names all risks (CAR) policy, in the names of the employer and the management contractor. In accordance with footnotes [h] and [k.1] cover under all risks policies will vary and the management contractor should agree any differences between the wording of all risks insurance in clause 6.2 and the policy presented. The management contractor also can use his own annual insurance policy as long as it is a joint names policy, under clause 6.4.3.1, and provided the cover is acceptable. Clause 6.4.8 provides a right of determination to either party, but this only applies to projects for the alteration or extension of existing structures.

Clause 6.5 concerns the insurance of existing structures when there are alterations and extensions to those structures. The employer is required to take out a joint names policy covering the specified perils. The management contractor for himself and for all works contractors must authorise the insurers to pay monies from such insurance to the employer. Clause 6.3 concerning cover for the works contractor under the employer's policy relieves the management contractor of taking out his own policy. Clause 6.5.3 provides the management contractor with the right to take out a joint names policy in the event of the employer failing to do so.

Clause 6.6 follows clause 22D of JCT 80 as amended in 1986. Thus it provides for the management contractor to take out insurance for the employer's loss of liquidated damages. However, the difference is that JCT 87 has no provisions similar to clause 25.4 of JCT 80 indicating the relevant events (see Chapter 14). In JCT 87 the relevant events including the event to which clause 22D refers are included in the Works Contract Conditions clause 2.10.

Clauses 6.7, 6.8 and 6.9 follow the wording of clauses 20.1, 20.1 and 20.3 respectively of JCT 80 as amended in 1986, reading 'Management Contractor' for 'Contractor', 'Project' for 'Works' and making the appropriate alterations in the clause numbers mentioned (see Chapter 3 for this wording).

Clause 6.10 follows the wording of clause 21 of JCT 80 as amended in 1986, except that the management contractor is required to 'cause any Works Contractor to take out and maintain insurance'. Appropriate alterations are made in the clause numbers mentioned (see Chapter 4 for this wording).

Clauses 6.11 and 6.12 follow the wording of clauses 21.2 and 21.3 of JCT 80 as amended in 1986 reading 'Management Contractor' for 'Contractor', 'Project' for 'Works' and appropriate alterations in the clause numbers mentioned (see Chapters 9 and 4 for this wording).

Works Contract/2 (Conditions)

Section 6 relates to injury, damage and insurance. Under clauses 6.2 and 6.3 the works contractor shall, in the case of personal injury or death (except where caused by the management contractor's or employer's fault), or damage to property (provided it is caused by the wrongdoing of the works contractor), indemnify the management contractor. The works contractor is also required to insure against claims the management contractor might make against him. The JCT management contract allows the employer to state in the appendix to the management contract the limit of indemnity of the public liability insurance which the works contractor must provide under clause 6.5.1 to cover any loss which a third party might sustain in connection with the works.

As regards responsibility for the works, the works contractor has the benefit of the management contractor's and employer's joint names policy for specified perils, in accordance with clause 6.4 etc. set out in the management contract JCT 87. The insurer has no right of subrogation against the works contractor, but the works contractor is likely to be responsible for the non-specified perils if they cause damage to the works.

18.2 The JCT Measured Term Contract 1989 Edition

This standard form came into use after its introduction in December 1989. Its object is to cater for employers who have a regular amount of maintenance work to be carried out on their premises. The injury, damage and insurance provisions follow the pattern of JCT 80 after the 1986 amendments. However, there are certain variations to conform with a measured term contract.

Section 6 deals with these provisions and the following are the main variations compared with JCT 1980 after the 1986 amendments, plus the necessary appropriate clause number amendments.

- Reference is made to 'an Order' instead of 'the Works'.
- There is no provision for partial possession by the employer.
- Reference is made to the 'Contract Administrator' instead of the 'Architect'.
- There is no provision for the clause 21.2.1 insurance regarding the subsidence risks for the employer.
- The employer must insure his existing structures and contents against the specified perils (not all risks) in joint names or waive subrogation rights against the *contractor*.
- In respect of the specified perils on existing structures and contents, the employer is required to notify his insurer of the properties to be covered by the contract, the contract period and the work or supply required.
- The all risks cover refers to 'work executed' or 'supplies'.
- There is no provision for a clause equivalent to 22D in JCT 1980 as amended in 1986 concerning liquidated damages.

Chapter 19

Contract guarantee and performance bonds

The Standard Form of Building Contract does not provide for sureties. However, as a form of security against the default of the contractor, a contract of guarantee or suretyship is frequently entered into, e.g. a performance bond. A bond is a promise by deed (thus it must be in writing, signed, sealed and delivered) whereby the person giving the promise (the obligor) promises to pay another (the obligee) a sum of money. When a guarantee is involved, the obligation determines upon the occurrence of the event guaranteed, e.g. where performance is guaranteed it is subject to a condition that it will come to an end upon performance of the contract. Section 130 of the Companies Act 1989 states that the insurance company's seal is no longer necessary and the document could be executed under the power of attorney route. This is useful for companies with more than one office as with only one seal allowed, it would be impossible to use any other method than power of attorney.

19.1 General points

The parties concerned and comparison with insurance

It is important to appreciate that a bond is *not* an insurance policy and that there are some basic differences between an insurance policy and a guarantee bond. An insurance policy is a contract between two parties only, the insured and the insurer, whereby in return for a premium the insurers agree to pay the claim should the event insured against occur.

A guarantee bond, however, involves three parties:

- the contractor, who is responsible for fulfilling the obligations set out in the bond. He has to perform some act under certain conditions or alternatively pay damages.
- The employer (the obligee), who is the beneficiary under the terms of a bond: either the obligation set forth is fulfilled or the amount of the bond is available so that the employer should be adequately protected.
- The surety (the obligor) is the party who joins with the contractor for the purpose of guaranteeing to the employer the fulfilment of the contractor's obligations.

Comparing this with insurance, the object of an employer in requiring a surety bond is protection against loss, and in that sense it is regarded as insurance. In suretyship, however, the contractor obtains no protection

254

himself under the bond. The employer, who is the third party to the contract, is the only one directly covered under a bond.

The basic difference between an insurance policy and a bond is that with an insurance policy the insurer takes the risk and the insured pays a premium for that privilege. With a bond the surety should dispose of all foreseeable risks before signing the bond and no responsibility passes from the contractor to his surety. The contractor still retains all his responsibilities in respect of his contract with the employer. In the event of a claim by the employer for non-performance, it is the contractor who has the responsibility of satisfying that non-performance, and only if he is unable to do so, will the surety's own assets be available.

An insured is under a duty to disclose to the insurer all facts material to the risk within the knowledge of the insured, but a surety can only complain of positive deception by representations expressed or implied. See *Seaton* v. *Heath* (1899), which was reversed without reference to this point under the name of *Seaton* v. *Bernard* (1900) and *Workington Harbour, etc.* v. *Trade Indemnity Co* (1933), which was also reversed on another point.

Characteristics of bonds

(1) By section 4 of the Statute of Frauds 1677

'No action shall be brought whereby to charge the defendant upon any special promise to answer for the debt, default, or miscarriage of another person ... unless the agreement upon which such action shall be brought ... shall be in writing and signed by the party to be charged therewith.'

Since the Mercantile Law Amendment Act 1856, it is not necessary for the consideration to appear in writing. However, in the case of bonds, no consideration is necessary. This is one reason why bonds are preferred to simple contracts of guarantee, because strictly speaking no consideration passes from the employer to the surety as ostensibly the contractor pays.

(2) An underlying obligation is necessary because a surety bond is not a contract complete in itself. There must be a contract existing between two of the parties to the bond, e.g. the contractor and the employer, before there can be a bond. The contractor has the same legal obligation to the employer that he would have if no bond were issued. His bond merely guarantees that he will perform the obligations he has undertaken.

(3) A bond is a joint and several obligation, it is not a one-party obligation, and once the bond is sealed by both contractor and his surety and delivered to the employer, the obligation is no longer that of the contractor alone. Upon default by the contractor, it is the surety's obligation as well. The employer may seek recovery from both the contractor and the surety or he may look only to the surety. Under English law the employer can sue either the surety for non-performance by the contractor or the contractor first. In the latter event, if unsuccessful, he can only add the cost of that first action to the sum

he ultimately claims against the surety provided he has first given the surety the opportunity to honour the guarantee in full. If the surety wishes the employer to sue the contractor first, he must guarantee to the employer all the costs of that first action.

(4) Under common law the contractor has numerous obligations to his surety, the most important of which is that he is required to discharge his obligation to the employer and in default the contractor must indemnify his surety for discharging the contractor's obligations in whole or in part. The contractor's common law obligations to his surety are normally reinforced by a counter indemnity (see further details under a separate heading later). In addition to those rights which the surety has under common law or which he secures from the contractor under the counter indemnity, the surety is subrogated to the right of the employer. In other words, if the surety pays a loss which the employer would otherwise suffer, the surety by operation of law acquires all the rights of the employer against the contractor to enforce performance of the contract or repayment of any losses paid by the surety. Thus all retention monies held by the employer are available to the surety.

Release by the employer

A surety bond remains in force until the obligation of the contractor has been fulfilled. It does not contain a cancellation clause as do many insurance policies, and in some cases does not even give a termination date. Once the obligations of the contractor have been completed to the satisfaction of the employer it is the employer who gives notification of the release of the bond, unless the bond, by its very wording, stipulates the event upon the happening of which the bond would terminate. For example, ICE Conditions include a form of bond which provides for its release at the date of the expiration of the period of maintenance.

The normal rule is that when the contractor has completed his performance the surety is discharged, unless the contract of guarantee requires certified completion. Where a building contract like the JCT form specifies stages of certified completion but the bond is silent as to which stage is meant, it is submitted that the final certificate should be taken as the stage of complete performance for the purposes of the bond (see Chapter 17, Section 17.6, under the heading 'Performance bond').

Banks or insurance companies for bonds?

The Insurance Companies Act 1982 states that only those insurance companies duly licensed by the Department of Trade and Industry under the relevant sections of the Act are allowed to provide performance bonds, the only exception being joint stock banks.

Contractors and employers have to consider whether their interests are better served by having a surety bond from a bank or an insurance company.

Private sureties can die, disappear or become bankrupt, which in the case of a bank or insurance company is extremely unlikely. In America, federal law precludes banks from being involved in domestic surety business. In Europe, banks are, generally speaking, allowed to transact domestic surety business.

When an insurance company issues a surety bond, it does not affect a contractor's banking lines of credit. There are, of course, occasions when only a banker's guarantee will be accepted by the employer, particularly with projects in certain Middle East countries. It is for these occasions that the contractor's bank bond facilities should be retained and utilised. If an insurer or bank is prepared to bond the contractor, it does not matter whether the employer is in the public or private sector. However insurers have a better prequalification process (see later) than banks.

For this reason and those mentioned later, public authorities, etc. now prefer the provision of the bond to be made by the insurance market where there are a small number of well established and experienced insurers prepared to underwrite the business.

The principal is relieved from making those searching enquiries which are necessary to decide if the contractor is suitable to carry out the work. It is true, of course, that the most thorough examination cannot exclude the possibility of default (large and well known firms have been forced into liquidation). Therefore, a performance bond issued by a reputable insurer is the accepted safe method by which a principal can be protected against loss. Thus, performance bonds are a vital safeguard to any principal against losses which inevitably follow the default of a contractor, and unfortunately the construction industry ranks high in the lists of insolvencies.

Whereas banks often issue a form of financial guarantee (which is an undertaking to pay on demand), by the terms of the insurance bond the insurer accepts responsibility for the performance of the obligations under the contract.

It is often simpler and quicker to obtain a bond from a bank because of their general knowledge of a client's affairs and because of their less complex underwriting approach to bonds, but banks are inclined to regard any bonds they issue as part of their customer's credit facilities. Insurers therefore argue that banks only treat the transaction as they would any financial transaction, such as an overdraft, i.e. they just want adequate security, whereas insurers carry out very thorough investigations all of which are of benefit to the employer as well as the insurer, and may at times assist the contractor. It follows, therefore, that any contractor who uses his bank for bond requirement does so at the expense of possible working capital, advances which the bank could otherwise make had it not got the bond commitment.

In conclusion with an insurance surety company the employer gets a prequalified contractor and a third party guarantee at the end of the day (if the contractor goes into liquidation for example), whereas a bank will just issue a bond provided it has sufficient collateral. Details of the insurers' investigations are given later under the heading 'Underwriting considerations'.

19.2 Performance bond wording and other details

On demand bonds – the *Perar* and *Trafalgar House* cases

Prior to the Court of Appeal decision in *Trafalgar House Construction (Regions) Ltd* v. *General Surety & Guarantee* (1994) it was thought that there was a sharp distinction to be drawn between on demand bonds (where the surety had to pay whatever the circumstances of the demand by the beneficiary, fraud excepted) and the more traditional conditional bonds (where the beneficiary must prove both the default and the damage he has suffered). Some 75 years of precedent had been overturned and, pending the House of Lords' decision in this case as to whether they agreed that conditional bonds on the present wording should be assimilated with on demand bonds on the grounds this is the commercial purpose lying behind the conditional bond, some drafters reworded their bonds to avoid this Court of Appeal decision. However, the Court of Appeal decision was reversed in the House of Lords where it was held that the bond was a guarantee and not an on demand instrument. Further details of this decision are given at the end of Appendix 3.

In *Perar* v. *General Surety & Guarantee* (1994) it was decided by the Court of Appeal that on the JCT 'automatic determination' clause 27.2 there was no breach of contract which could trigger the bondsman's liability. It was not argued that this conditional bond in similar form to that considered in the *Trafalgar House* case was in effect an on demand bond. The point in this case was that, to establish liability under conditional bonds, there must be a default on the part of the contractor. In the case of automatic determination, the default does not arise on or by reason of the automatic determination following receivership. However, it will arise following a failure to pay by the contractor in receivership, when the ascertainment procedure (e.g. the cost of engaging and paying others to complete the works, and a final accounting and a certificate indicating a balance one way or the other) has been completed under clause 27.4. This means that the receiver (almost inevitably) refuses or fails to pay any balance due. At that stage the surety will be liable.

To some extent the *Perar* case illustrates indirectly the fact that, on the traditional conditional bond wording (as interpreted prior to the Court of Appeal decision in *Trafalgar House*), the beneficiary has to wait for his money from the surety, and this was one aspect that the Court of Appeal in the latter case did not like. The court felt that the beneficiary wants some assurance of immediate funds and, as the final accounting position between the parties to the construction contract is likely to be a long and laborious process in many cases, the archaic wording allowed the court to confine the payment to the damages suffered by the beneficiary. These damages were limited to the additional expenditure incurred or to be incurred by the beneficiary as a result of the failure to perform the contract.

The other and more popular view is that an on demand bond is subject to considerable abuse, as the surety is in no position to demand proof of loss or even that a default has occurred, and is therefore to be avoided.

In the *Trafalgar House* case the Court of Appeal said that, provided the demand stated the amount of damages and was made bona fide, there would be immediate liability in that amount and the presence of cross-claims of any kind was irrelevant. As a result, all surety insurance companies felt that they must amend their bond wordings to make clear what they consider to be the correct coverage notwithstanding the outcome of the House of Lords decision on the *Trafalgar House* case. There never has been, nor was it ever considered to be the case, that surety bonds are issued as independent financial guarantees without any respect to the underlying contract they support. Insurers therefore believe it reasonable that, before paying out under a claim like any other form of insurance, they are entitled to receive proof that a loss has been suffered and that the loss be duly quantified. Surety bonds are no exception.

The ABI model for discussion

Following the Latham Report the Department of the Environment has commissioned solicitors Ashurst Morris Crisp to produce a report which hopefully will lead to the creation of a new model performance bond wording to be used by all Government departments, local authorities and the like. The Association of British Insurers and these solicitors have produced a model for discussion (see Appendix 3).

The following comments concern the wording of the latest amended model form of performance bond.

(1) The first section sets out the three parties to this bond.

(2) The second section identifies the construction contract concerned in the schedule and makes the contract a part of the bond; and ensures that the guarantor has agreed with the employer to guarantee the performance of the contractor under the contract subject to the limitation of the bond amount set out in the schedule.

(3) The third section consists of seven terms or conditions as follows:
(a) Subject to the bond provisions the guarantor guarantees to the employer the performance of the contractor's obligations contained in the contract. In the event of a breach of the contract the guarantor will satisfy the damages sustained by the employer in accordance with the provisions of the contract and taking into account all sums due or to become due to the contractor. This latter point was ignored in the *Trafalgar House* decision in the Court of Appeal.
(b) The maximum aggregate liability of the guarantor and the contractor

under this bond is the bond amount, but otherwise the terms and conditions of the contract apply in determining the liability of the guarantor as apply in deciding the liability of the contractor.

(c) This is the old indulgence (or forbearance) clause in a modern wording. It gives the employer the right to alter any of the terms and conditions of the contract relating to the extent or nature of the works and allowance of time by the employer, without releasing, reducing or affecting the liability of the guarantor under the bond. This avoids the guarantor's common law right.

(d) The guarantor is released in accordance with the expiry date in the schedule, subject to any breach of the contract of which notice in writing has been given to the guarantor before the expiry date.

(e) The contractor undertakes to the guarantor, without limiting other rights and remedies of the employer or the guarantor against the contractor, to perform the contractual obligations.

(f) There shall be no assignment of the bond without the prior written consent of the guarantor and the contractor.

(g) The laws of England, Wales and Scotland shall apply as chosen.

(4) This fourth section is the schedule, which includes the names and addresses of the parties to the bond, details of the contract between the employer and the contractor, the bond amount and the expiry date.

(5) Finally there is an attestation clause which provides for the seals of both the contractor and the guarantor.

The draft Model Form of Guarantee Bond distributed for discussion and subsequently amended makes it clear that the obligations of the guarantor relate to the underlying contract and that the guarantor is liable to the extent that the contractor himself is liable for any breach of contract, subject to the bond amount. The ABI invited comments upon the draft form from the construction industry and all other interested parties before 30 December 1994. This resulted in the subsequent amendment now set out in Appendix 3, and published in September 1995.

The amount and cost of a bond

Although the normal amount of a performance bond is 10% of the contract price, there is a trend to increase this figure as it is doubtful whether 10% is sufficient. Perhaps 20% may become more usual.

In some overseas countries a 100% bond may be required, although insurers are usually reluctant to agree to these.

The cost of performance bonds is relatively low, although it does vary according to circumstances. The stability and experience of the contractor are of prime importance, but other features may affect rating, e.g. bond percentage, the size of retention, and period of contract.

As with all insurance contracts, the underwriter assesses the risk and he will charge a higher premium for those risks which he considers to be more hazardous.

19.3 Putting a bond in place

Underwriting considerations

The following aspects are investigated:

- the technical ability of the contractor;
- the usual type of contract undertaken by the contractor;
- the contractor's tender amount;
- a progress report on the contracts already bonded by the insurer for the contractor;
- whether the contractor's business is well run and successful.

The whole financial position will be considered: the extent of the contractor's indebtedness and whether overdrafts and loans are reasonable in relation to the size of the business. The amount of liquidity is vital.

In the case of a first application a standard questionnaire is completed by the contractor and the last three years' audited accounts are made available to the insurers. If the contractor is a member of a group of companies, the same information is required from the group.

The object of these investigations is to discover whether the contractor satisfies the following criteria, i.e. whether the contracting firm:

- is run competently;
- has a record of completing contracts from the technical and profitability viewpoints as well as having adequate resources of manpower, plant and equipment;
- has sufficient financial resources to carry out the operations of the business after considering the possibility of unknown future contingencies.

It should be noted that bond underwriters do not only make decisions on historical information. They also wish to look at current management information, the strength of reporting systems, cash flow forecasts, bank status reports and up-to-the-minute cost value reconciliations.

Many contractors in the construction industry overtrade: their financial resources may be inadequate to meet their probable commitments. With luck they survive but if, for example, they meet with unexpected problems on a contract or they are faced with unanticipated higher interest charges on their loans, they may encounter serious cash flow problems and at the worst be unable to carry on their business. Thus contractors fail because they run out of cash, not because they record a loss in their profit and loss account.

Counter indemnities

Although there is a common law right of indemnity from the bonded contractor in favour of the surety company, there is a growing practice among sureties to require a specific indemnity form to be sealed by the bonded contractor. A counter indemnity should always be required as a matter of course from the ultimate parent company of the contractor. A wording of a form of counter indemnity (or guarantee) is given in Appendix 4.

It is important, in the case where the contractor is a subsidiary company, that the insurer surety should obtain a counter guarantee from the holding company, especially as the subsidiary is usually a limited liability company and consequently a separate legal entity. The main reason is that it cannot be assumed that the holding company will support the subsidiary, as it may be economically sound for the parent company to allow the subsidiary to go into liquidation rather than continue to incur losses. Where a proper counter guarantee is taken and the contractor encounters difficulties, the holding company will either have to support the subsidiary financially or reimburse the surety and, in order to keep the reimbursement figure to a minimum, the parent company will no doubt assist in the completion of the contract with the least possible delay. If the parent company will not provide the counter guarantee required, the insurer will have to decide whether the risk is acceptable on any other basis, e.g. a form of collateral security.

Where there is not just a parent company and a subsidiary but a group of companies, sureties are no longer prepared to rely upon an indemnity given by the parent or holding company alone. They have now joined the ranks of the bankers in this regard and they take cross guarantees, i.e. from all companies within the group.

Additionally, where the directors have substantial personal share-holdings, or where a substantial part of a contractor's capital comprises directors' loans or undrawn remuneration, or even simply low paid up capital, the surety should require each director to sign a personal guarantee.

In essence, a counter indemnity is a written undertaking given by the contractor or some other party in consideration of the surety issuing a bond to the principal to repay to the surety all losses, costs, charges and expenses which the surety may be called upon to pay or discharge under or by virtue of the bond to be issued. Where personal guarantors are required, the signatories accept personal responsibility to fulfil the terms of the guarantee.

Most counter indemnities require the reimbursement of monies which the insurer has paid in respect of its obligations under the bond. On the other hand, some insurers may attempt to introduce a form of 'on demand' wording which requires reimbursement before the insurers have actually made any payment. However, the case of *General Surety & Guarantee Co Ltd* v. *Francis Parker Ltd* (1977) makes it clear that the wording has to be precise before there is a liability to pay on demand. The court commented that a simple way of dealing with this matter is to provide that, as between the contractor and the insurer, a demand by a principal should be conclusive evidence of the insurer's liability to the principal. Therefore the tendency is for insurers to have a tighter indemnity agreement.

Acceptability

The insurers, in practice, will set their standards for the acceptance of risks, and regrettably many applications will prove to be unacceptable. Sometimes the decision can be changed by the offer of additional security from the contractor. A charge on the business is not usually possible because the bank will already have this as a security for the overdraft. However, the directors or principals (or their wives) often have personal assets, property or land, which are unencumbered, and these can be offered as a collateral security. Sometimes the equity in the principal's own private dwellinghouse may be offered, but this is less than satisfactory. If it is acceptable, it is necessary for the principal's wife to join in the guarantee.

Occasionally a substantial cash deposit may be available. However, this has the effect of reducing the contractor's working capital and can only exacerbate his difficulties. If one insurer records a declinature, the contractor is forced to try other underwriters or he may persuade his bank to offer assistance.

If a bond cannot be secured it means that there is a serious doubt about the contractor's ability to complete the contract and the employer must either award the contract to another or take the risky course of allowing the contractor to proceed without a bond being in force.

19.4 Types of bond

Bulk schemes

Sometimes insurers come to an agreement with a local authority or other regular employer of contractors to have the first right to provide the bond required from any contractor whose tender is accepted by the employer. In return the insurer quotes a special advantageous rate of premium.

At one time, it was thought that agreements such as these bulk contract guarantee schemes would have to be registered under the Fair Trading Act 1973. However, although it is true that the contractor has no freedom of choice of insurer, as already mentioned, he is not in fact paying for the bond. The real parties to the agreement are the employer and the insurer. The employer clearly has a choice and need not enter into such an agreement unless it is to his advantage, i.e. he obtains a very satisfactory rate of premium. In return the insurer gets bulk business. It is also arguable that public and local authorities are not strictly speaking 'persons who carry on business' within the meaning of the 1973 Act.

On 2 July 1976 the district and county councils associations issued a directive to their members as follows:

'If the employer nominates the bond guarantor in the tender documents, provision should be made for alternative sureties to be put forward by the tenderer for the approval of the employer. Such approval should not be unreasonably withheld.'

Certain local authorities and some large private sector employers now nominate their own bond consultants in tender documents. The successful contractor must arrange his bond for that contract through the nominated consultants but normally such schemes allow any surety acceptable to the employer to be used.

On demand bonds

Bonds issued in this country usually require proof of the amount of the loss incurred before any payment can be made. However, in connection with overseas bonds it is common for an 'on demand' wording to be employed. In other words, in the event of the contract not proceeding for some reason or the contractor being in default, the principal can demand payment of the full bond amount without giving any proof of the amount of loss or indeed that any loss has occurred at all.

In *Edward Owen Engineering Ltd* v. *Barclays Bank International Ltd* (1977) an English company contracted to erect glasshouses in Libya. The contract never started because of difficulties regarding the letter of credit arranged by the Libyan customers for the payment of instalments. However, a guarantee had been given by the bank on behalf of the contractors. The bank had said, 'we confirm our guarantee ... payable on demand without proof or conditions'. A counter guarantee had been obtained by the bank from the contractors, and it was the contractors who brought the action in an attempt to prevent Barclays Bank from paying under the bond. The Court of Appeal held that in the absence of fraud the bank had to fulfil its duty to honour the guarantee it had issued.

Insurers are generally reluctant to accept these 'on-demand' wordings, for it is only right that any claim shall be investigated properly and an amount paid which represents the real loss to the principal. Hence the opposition to the *Trafalgar House* Court of Appeal decision mentioned earlier, as on demand bonds are truly independent pure financial guarantees, tantamount to irrevocable letters of credit.

19.5 How claims arise

A contractor may, because of financial problems, be forced to stop work on a site. A receiver may be appointed on behalf of creditors or a liquidator may be appointed to wind up the company. Occasionally, however, difficulties may arise between the contractor and the principal which lead to the withdrawal of the contractor from the site. The contractor may remain in business but in such an event a court action will probably be necessary to decide the liabilities of the respective parties.

When a receiver is appointed, his main duty will be to look after the interest of the creditors. There is nothing to prevent him from carrying on the business, and if a contract is likely to be profitable, it will be continued and completed. There is, however, usually little incentive to complete a contract which has lost and will continue to lose money.

The principal will be anxious that the contract should be completed on time with the minimum of additional cost or trouble. When the original contractor has stopped work, it is usually necessary for fresh tenders to be obtained for completion of the outstanding work. There are often difficulties in connection with completed work which may not be up to the required standard. It is inevitably much more costly for new contractors to come on to a site to complete and/or repair someone else's work.

Underwriters, when assessing the risk, pay particular attention to the level of other tenders originally obtained. If there are numerous tenders, fairly close together, it can be assumed there will not be much difficulty in finding other contractors to complete in the event of default. If, however, there are few tenders with wide variations, underwriters will ask many more questions. Often, with specialist work, there are just one or two contractors who are interested in tendering for contracts: similarly, in isolated areas, the number of potential contractors may be limited. These problems can all increase the likely cost of a claim in the event of a default.

As soon as they are aware of the possibility of a claim, sureties will usually appoint engineers or surveyors to protect their interests. The amount and quality of work completed will have to be assessed, together with the work still to be done. The interests of principals and sureties do not necessarily coincide: sometimes principals do not wish to put the work out to tender or to accept the lowest tender available.

Where there are counter guarantors involved (for example, where personal guarantees are obtained from directors), there is a further possible conflict of interest. The counter guarantors may not agree with the actions taken by either the principals or the insurers: they may, for example, disagree about the extent of defects in work already done.

The protection of the site is important during the period between the cessation and recommencement of work. The principals have a duty to ensure the safety of already completed work and the materials on the site. Arrangements have to be made for the CAR policy to be maintained until fresh cover is effected by the new contractor. In practice the liquidator or receiver may arrange cover and accept responsibility for the premium. Uncompleted sites are especially liable to thefts and damage by vandals, and it is in the interests of all that effective security arrangements be implemented. The principals are likely to add the cost of these to the claim being made.

It is unusual for the amount of any retention money held to be sufficient to pay the additional costs involved in completing the work. There is an inevitable time lag, and inflation alone tends to increase the costs. It is not unusual for the whole of the bond amount to be swallowed up, especially when the original tender price was low.

After paying a claim, the insurers are entitled to recover under any counter indemnity they may have or to realise any collateral security.

Unfortunately personal guarantees are often of little value, as these may have been given to all and sundry. There may be no money available or it may be effectively tied up in other names or other countries so that recovery is not possible.

The sureties will rank as ordinary creditors in the winding up of the company, but the contractor's bankers will usually have preference, as will the Inland Revenue. The amount available for distribution to other creditors is often very small.

What often happens is that sureties providing large facilities, certainly to all national contractors, will not be using a standard indemnity but one which enables them to rank *pari passu* with the bank in terms of security, i.e. the indemnity provides for a negative pledge clause. In simple terms this means that the surety accepts that certain security has been passed to the bank, e.g. specific charges on the assets of the business, but should further security be passed or the bank make a move for a fixed or floating charge, i.e. a debenture, then the surety shares *pari passu* in the additional security so afforded.

19.6 Bonding of sub-contractors

Bonding of sub-contractors has to be considered both from the point of view of the main contractor in connection with either domestic or nominated sub-contractors and also from that of the employer – but only with regard to nominated sub-contractors. The main contractor is responsible for completing the works and if one of his domestic sub-contractors fails, the main contractor has to solve the problem at his own expense. It is, however, becoming increasingly common for main contractors to insist on bonds from their own domestic sub-contractors, particularly where the sub-contract amounts are substantial. The position with regard to nominated sub-contractors is far more complex and one has to refer to the main contract document to establish the responsibilities of the various parties in the event of the failure of a nominated sub-contractor.

On the insolvency of a nominated sub-contractor under sub-clause 35.24.5 of JCT 80 'the architect shall make such further nomination of a sub-contractor ... as may be necessary'. This follows the decision in *Bickerton* v. *North West Metropolitan Hospital Board* (1969), and in accordance with that case the presumption is that the employer must pay for the cost of this re-nomination under JCT 80.

19.7 Other kinds of bond for the construction industry

The various other types of bond, which can be required at any stage in a contract, will now be considered.

(1) *Bid or tender bonds* are intended to assure the buyer or the employer that the bid is a responsible one and that if the tenderer's bid is accepted they will

proceed and effect the form of contract including any necessary subsequent bonding requirements. If the tenderer fails to do so, the losses of the buyer would form a claim under the bond to be met by the tenderer or his surety.

(2) *Advance payment, progress or repayment bonds* are similar to performance bonds in that the buyer receives a guarantee that any monies advanced will not be lost through default or poor performance by the contractor.

(3) *Maintenance bonds* are normally requested in connection with construction contracts. They guarantee that once the construction has been completed the contractor will fulfil his obligations throughout the maintenance period and may be in lieu of retention monies during the maintenance period.

(4) *Retention bonds* are given in lieu of retention monies required throughout the contract period or against the early release of the retention element of monthly progress payments.

(5) Highways Act bonds guarantee the completion of street works and usually are 100% of the estimated works cost.

All the various classes of bond mentioned can be worded to be either conditional or on demand.

Appendix 1

Specimen 'special perils' extension

SPECIAL PERILS – MATERIAL DAMAGE

Use Part A of the Master Wording and incorporate the Special Conditions from Part B as indicated in Part C.

MASTER WORDING

Part A

The insurance by (item(s) ... of) this policy shall subject to all the Exclusions Provisions and Conditions of this policy (except in so far as they may be hereby expressly varied) and the Special Conditions set out below extend to include:

> Here insert the wording(s) from the boxes in Part C appropriate to the peril(s) selected

Part B

SPECIAL CONDITIONS

1 The liability of the Insurer under this extension in respect of any item of the policy not subject to any condition of Average (Underinsurance) shall be limited to the proportion which the sum insured by such item shall bear to the total insurances effected by or on behalf of the Insured on the same property against DAMAGE by fire as shown on the face of this policy.

2 Each item of this policy which is subject to any condition or conditions of Average (Underinsurance) is subject to the same condition or conditions under this extension in like manner. The liability of the Insurer under this extension in respect of each item of this policy shall be limited to the proportion which the sum insured thereunder shall bear to the total insurances effected by or on behalf of the Insured on the same property against DAMAGE by fire as shown on the face of this policy.

3 The liability of the Insurer under this extension and the policy in respect of an item shall not exceed the sum insured by such item.

4 This insurance does not cover
   (a) DAMAGE occasioned by
       (i) riot or civil commotion (except to the extent they may be specifically insured hereby)

    (ii) war invasion act of foreign enemy hostilities (whether war be declared or not) civil war rebellion revolution insurrection or military or usurped power

(b) loss or destruction of or damage to any property whatsoever or any loss or expense whatsoever resulting or arising therefrom directly or indirectly caused by or contributed to by or arising from

    (i) ionising radiations or contamination by radioactivity from any nuclear waste from the combustion of nuclear fuel

    (ii) the radioactive toxic explosive or other hazardous properties of any explosive nuclear assembly or nuclear component thereof

(c) loss or destruction or damage caused by pollution or contamination but this shall not exclude destruction of or damage to the Property Insured, not otherwise excluded, caused by

    (i) pollution or contamination which itself results from a peril hereby insured against

    (ii) any peril insured against which itself results from pollution or contamination

(d) consequential loss or damage of any kind or description except loss of rent when such loss is included in the cover under this policy

(e) DAMAGE in Northern Ireland occasioned by or happening through or in consequence of

    (i) civil commotion

    (ii) any unlawful wanton or malicious act committed maliciously by a person or persons acting on behalf of or in connection with any unlawful association

    For the purposes of this Condition

      "unlawful association" means any organisation which is engaged in terrorism and includes an organisation which at any relevant time is a proscribed organisation within the meaning of the Northern Ireland (Emergency Provisions) Act 1973

      "terrorism" means the use of violence for political ends and includes any use of violence for the purpose of putting the public or any section of the public in fear

    In any action, suit or other proceedings where the Insurer alleges that by reason of the provisions of this Condition any DAMAGE is not covered by this policy, the burden of proving that such DAMAGE is covered shall be upon the Insured

5   The Insured shall take all reasonable precautions to prevent DAMAGE.

6   On the happening of any DAMAGE from peril(s) ... above full details thereof shall be given to the Insurer within seven days.

7   In so far as this insurance relates to DAMAGE (other than by fire or explosion) directly caused by Malicious Persons, it is a condition precedent to any claim that immediate notice of the DAMAGE shall have been given by the Insured to the Police Authority.

8   In so far as this insurance relates to DAMAGE caused by Subsidence Ground Heave or Landslip

(a) The Insured shall notify the Insurer immediately they become aware of any demolition, groundworks, excavation or construction being carried out on any adjoining site;

(b) The Insurer shall then have the right to vary the terms or cancel this cover.

Part C

1 EXPLOSION

DAMAGE caused by EXPLOSION excluding DAMAGE

(a) caused by or consisting of the bursting of a boiler (not being a boiler used for domestic purposes only) economiser or other vessel machine or apparatus in which internal pressure is due to steam only and belonging to or under the control of the Insured

(b) in respect of and originating in any vessel machinery or apparatus, or its contents, belonging to or under the control of the Insured which requires to be examined to comply with any Statutory Regulations unless such vessel machinery or apparatus shall be the subject of a policy or other contract providing the required inspection service

(c) by pressure waves caused by aircraft or other aerial devices travelling at sonic or supersonic speeds

and Special Conditions 1 and 3 in Part B

2 AIRCRAFT

DAMAGE (by fire or otherwise) caused by AIRCRAFT or other aerial devices or articles dropped therefrom excluding DAMAGE by pressure waves caused by aircraft or other aerial devices travelling at sonic or supersonic speeds

and Special Conditions 1 and 3 in part B

3 RIOT

DAMAGE (by fire or otherwise including explosion) caused by RIOT CIVIL COMMOTION STRIKERS LOCKED-OUT WORKERS or persons taking part in labour disturbance or malicious persons acting on behalf of or in connection with any political organisation excluding DAMAGE

(a) arising from confiscation requisition or destruction by order of the government or any public authority

(b) arising from cessation of work

and Special Conditions 1, 3 and 6 in Part B

4 RIOT (FIRE ONLY)

DAMAGE (by fire only) caused by RIOT or CIVIL COMMOTION excluding DAMAGE arising from
(a) confiscation requisition or destruction by order of the government or any public authority
(b) cessation of work

and Special Conditions 1, 3 and 6 in Part B with Special Condition 4 (a)(ii) amended by replacing the words in brackets with "(other than by fire)"

5 RIOT AND MALICIOUS DAMAGE

DAMAGE caused by RIOT CIVIL COMMOTION STRIKERS LOCKED-OUT WORKERS or persons taking part in labour disturbances or MALICIOUS PERSONS excluding
(a) DAMAGE arising from confiscation requisition or destruction by order of the government or any public authority,
(b) DAMAGE arising from cessation of work,
(c) as regards DAMAGE (other than fire or explosion) directly caused by Malicious Persons not acting on behalf of or in connection with any political organisation
  (i)   DAMAGE by theft
  (ii) DAMAGE in respect of any building which is empty or not in use
  (iii) the first of £   of each and every loss as ascertained after the application of any condition of Average (Underinsurance)

and Special Conditions 1, 3 and 6 in Part B.

6   (a) EARTHQUAKE – FIRE RISK ONLY

DAMAGE (by fire only) caused by EARTHQUAKE

(b) EARTHQUAKE – SHOCK RISK ONLY

DAMAGE caused by EARTHQUAKE excluding DAMAGE by fire

(c) EARTHQUAKE – FIRE AND SHOCK RISKS

DAMAGE caused by EARTHQUAKE

and Special Conditions 1 and 3 in Part B

10 STORM AND FLOOD

DAMAGE caused by STORM OR FLOOD excluding
(a) DAMAGE attributable solely to change in the water table level
(b) DAMAGE by frost subsidence ground heave or landslip
(c) DAMAGE in respect of movable property in the open, fences and gates
(d) the first £   of each and every loss in respect of each separate premises as ascertained after the application of any condition of Average (Under-insurance)

and Special Conditions 1 and 3 in Part B

11 ESCAPE OF WATER

DAMAGE caused by ESCAPE OF WATER FROM ANY TANK APPARATUS OR PIPE excluding
(a) DAMAGE by water discharged or leaking from any automatic sprinkler installation
(b) DAMAGE in respect of any building which is empty or not in use
(c) the first £   of each and every loss in respect of each separate premises as ascertained after the application of any condition of Average (Under-insurance)

and Special Conditions 1 and 3 in part B

Appendix 2

Specimen contractors'
combined policy

Contractors' Insurance

In consideration of the payment of the premium the Independent Insurance Company Limited (the Company) will indemnify the Insured in the terms of this Policy against the events set out in the Sections operative (specified in the Schedule) and occurring in connection with the Business during the Period of Insurance or any subsequent period for which the Company agrees to accept payment of premium.

The Proposal made by the Insured is the basis of and forms part of this Policy.

M.J. Bright
Chairman and Managing Director

Independent Insurance Company Limited,
Administrative Office and Registered Office,
Marsland House,
Marsland Road,
Sale,
Cheshire M33 3AQ.

Registered Number 80623, England.

Definitions

1. **Proposal** shall mean any information provided by the Insured in connection with this insurance and any declaration made in connection therewith.

2. **Business** shall include

   (a) the provision and management of canteens clubs sports athletics social and welfare organisations for the benefit of the Insured's Employees

   (b) the ownership repair maintenance and decoration of the Insured's premises and the provision and management of first aid fire and ambulance services

   (c) private work carried out by an Employee of the Insured (with the consent of the Insured) for any director partner or senior official of the Insured.

3. **Employee** shall mean

   (a) any person under a contract of service or apprenticeship with the Insured

   (b) i) any labour master or labour only subcontractor or person supplied or employed by them

   ii) any self-employed person

   iii) any person hired or borrowed by the Insured from another employer under an agreement by which the person is deemed to be employed by the Insured

   iv) any student or person undertaking work for the Insured under a work experience or similar scheme

   while engaged in the course of the Business.

4. **Bodily Injury** shall include

   (a) death illness or disease

   (b) wrongful arrest wrongful detention false imprisonment or malicious prosecution

   (c) mental injury mental anguish or shock but not defamation.

5. **Damage** shall mean loss of or damage.

6. **Property** shall mean material property.

7. **Territorial Limits** shall mean

   (a) Great Britain Northern Ireland the Isle of Man the Channel Islands or off shore installations within the continental shelf around those countries

   (b) member countries of the European Economic Community where the Insured or directors partners or Employees of the Insured who are ordinarily resident in (a) above are temporarily engaged on the Business of the Insured

   (c) elsewhere in the world where the Insured or directors partners or Employees of the Insured who are ordinarily resident in (a) above are on a temporary visit for the purpose of non-manual work on the Business of the Insured.

8. **Excess** shall mean the total amount payable by the Insured or any other person entitled to indemnity in respect of any Damage to Property or the Property Insured arising out of any one event or a series of events arising out of one original cause before the Company shall be liable to make any payment.

   If any payment made by the Company shall include the amount for which the Insured or any other person entitled to indemnity is responsible such amount shall be repaid the Company forthwith.

9. **Contractual Liability** shall mean liability which attaches by virtue of a contract or agreement but which would not have attached in the absence of such contract or agreement.

10. **Contract Works** means the temporary or permanent works executed or in course of execution by or on behalf of the Insured in the development of any building or site or the performance of any contract including materials supplied by reason of the contract and other materials for use in connection therewith.

11. **Principal** shall mean any person firm company ministry or authority for whom the Insured is undertaking work.

Section 1 – Employer's Liability

In the event of Bodily Injury caused to an Employee within the Territorial Limits the Company will indemnify the Insured in respect of all sums which the Insured shall be legally liable to pay as compensation for such Bodily Injury arising out of such event.

Avoidance of Certain Terms and Right of Recovery

The indemnity provided under this Section is deemed to be in accordance with such provisions as any law relating to the compulsory insurance of liability to Employees in Great Britain Northern Ireland the Isle of Man or the Channel Islands may require but the Insured shall repay to the Company all sums paid by the Company which the Company would not have been liable to pay but for the provisions of such law.

World-wide

The indemnity granted by this Section extends to include liability for Bodily Injury caused to an Employee whilst temporarily engaged in manual work outside the Territorial Limits

Provided that

(a)   such Employee is ordinarily resident within Great Britain Northern Ireland the Isle of Man or the Channel Islands

(b)   the Company shall not be liable to indemnify the Insured in respect of any amount payable under Workmen's Compensation Social Security or Health Insurance legislation.

Section 2 – Public Liability

In the event of accidental

(a)   Bodily Injury to any person

(b)   Damage to Property

(c)   obstruction trespass or nuisance

occurring within the Territorial Limits the Company will indemnify the Insured in respect of all sums which the Insured shall be legally liable to pay as compensation in respect of such event.

The Company shall not be liable for any amount exceeding the Limit of Indemnity.

Motor Contingent Liability

Notwithstanding Exception 2(c) below the Company will indemnify the Insured within the terms of this Section in respect of liability for Bodily Injury or Damage to Property caused by or through or in connection with any motor vehicle or trailer attached thereto (not belonging to or provided by the Insured) being used in the course of the Business

Provided that the Company shall not be liable for

(a)   damage to any such vehicle or trailer

(b)   any claim arising whilst the vehicle or trailer is

  i)   engaged in racing pacemaking reliability trials or speed testing

  ii)   being driven by the Insured

  iii)   being driven with the general consent of the Insured or of his representative by any person who to the knowledge of the Insured or other such representative does not hold a licence to drive such a vehicle unless such a person has held and is not disqualified from holding or obtaining such a licence

  iv)   used elsewhere than in Great Britain Northern Ireland the Isle of Man or the Channel Islands.

Defective Premises Act 1972

The indemnity provided by this Section shall extend to include liability arising under Section 3 of the Defective Premises Act 1972 or Section 5 of the Defective Premises (Northern Ireland) Order 1975 in respect of the disposal of any premises which were occupied or owned by the Insured in connection with the Business

Provided that the Company shall not be liable for the cost of remedying any defect or alleged defect in such premises.

Movement of Obstructing Vehicles

Exception 2(c) shall not apply to liability arising from any vehicle (not owned or hired or lent to the Insured) being driven by the Insured or by any Employee with the Insured's permission whilst such vehicle is being moved for the purpose of allowing free movement of any vehicle owned hired by or lent to the Insured or any Employee of the Insured

Provided that

(a)   movements are limited to vehicles parked on or obstructing the Insured's own premises or at any site at which the Insured are working

(b)   the vehicle causing obstruction will not be driven by any person unless such person is competent to drive the vehicle

(c)   the vehicle causing obstruction is driven by use of the owner's ignition key

(d)   the Company shall not indemnify the Insured against

  i)   damage to such vehicle

  ii)   liability for which compulsory insurance or security is required under any legislation governing the use of the vehicle.

Leased or Rented Premises

Exception 4(b) shall not apply to Damage to premises leased or rented to the Insured

Provided that the Company shall not indemnify the Insured against

(a)   Contractual Liability

(b)   the first £100 of Damage caused otherwise than by fire or explosion.

Exceptions

The Company shall not indemnify the Insured against liability

1.   in respect of Bodily Injury to any Employee arising out of and in the course of his employment by the Insured.

2.   arising out of the ownership possession or use by or on behalf of the Insured of any

  (a)   aircraft aerospatial device or hovercraft

  (b)   watercraft other than hand propelled watercraft or other watercraft not exceeding 20ft in length

  (c)   mechanically propelled vehicle licensed for road use including trailer attached thereto other than liability caused by or arising out of

    i)   the use of plant as a tool of trade on site or at the premises of the Insured

    ii)   the loading or unloading of such vehicle

    iii)   damage to any building bridge weighbridge road or to anything beneath caused by vibration or by the weight of such vehicle or its load

Section 2 – Public Liability

but this indemnity shall not apply if in respect of such liability compulsory insurance or security is required under any legislation governing the use of vehicle.

3.   for Damage to Property which comprises the Contract Works in respect of any contract entered into by the Insured and occurring before practical completion or a certificate of completion has been issued.

4.   in respect of Damage to Property

   (a)   belonging to the Insured

   (b)   in the custody or under the control of the Insured or any Employee (other than Property belonging to visitors directors partners or Employees of the Insured)

   Exception 4(b) shall not apply to Damage to buildings (including contents therein) which are not owned or leased or rented by the Insured but are temporarily occupied by the Insured for the purpose of maintenance alteration extension installation or repair.

5.   for the cost of and expenses incurred in replacing or making good faulty defective or incorrect

   (a)   workmanship

   (b)   design or specification

   (c)   materials goods or other property supplied installed or erected

   by or on behalf of the Insured but this Exception shall not apply to accidental Damage which occurs as a direct consequence to the remainder of the Contract Works which are free of such fault defect or error.

6.   caused by or arising from advice design or specification provided by or on behalf of the Insured for a fee.

7.   for the Excess specified in the Schedule other than for Damage to premises leased or rented by the Insured.

8.   caused by or arising from seepage pollution or contamination unless due to a sudden unintended and unexpected event.

Use of heat

It is a condition precedent to the liability of the Company that when

(a)   welding or flame-cutting equipment blow lamps blow torches or hot air guns are used by the Insured or any Employee away from the Insured's premises the Insured shall ensure that

   i)   all moveable combustible materials are removed from the vicinity of the work

   ii)   suitable portable fire extinguishing apparatus will be kept ready for immediate use as near as practicable to the scene of the work

   iii)   before heat is applied to any wall or partition or to any material built into or passing through a wall or partition an inspection will be made prior to commencement of each period of work to make certain that there are no combustible materials which may be ignited by direct or conducted heat on the other side of the wall or partition

   iv)   they are lit as short a time as possible before use and extinguished immediately after use and that they are not left unattended whilst alight

   v)   blow lamps are filled and gas cylinders or cannisters are changed in the open

   vi)   the area in which welding or flame-cutting equipment is used will be screened by the use of blankets or screens of incombustible material

   vii)   a fire safety check is made in the vicinity of the work on completion of each period of work

(b)   vessels for the heating of asphalt or bitumen are used away from the Insured's premises the Insured shall ensure that each vessel

   i)   shall be kept in the open whilst heating is taking place

   ii)   shall not be left unattended whilst heating is taking place

   iii)   if used on a roof shall be placed upon a surface of non-combustible material

   iv)   shall be suitable for the purpose for which it is intended and be maintained and used strictly in accordance with the manufacturer's instructions.

Property in the Ground

The indemnity provided by this Section shall not apply to liability in respect of Damage to pipes cables mains and other underground services unless the Insured

1.   has taken or caused to be taken all reasonable measures to identify the location of pipes cables mains and other underground services before any work is commenced which may involve a risk of Damage thereto

2.   has retained a written record of the measures which were takn to comply with 1. above before such work has commenced

3.   has adopted or caused to be adopted a method of work which minimises the risk of Damage to such pipes cables mains and other underground services.

Section 3 – Contract Works

In the event of Damage to the Property Insured the Company will by payment or at its option by repair reinstatement or replacement indemnify the Insured against such Damage

Provided that

1. the Company shall not indemnify the Insured

    (a) under Items 1, 2, 3 and 5 of the Property Insured for any amount exceeding the Limit of Indemnity in respect of each Item in any one Period of Insurance

    (b) under Item 4 of the Property Insured for any amount exceeding the Limit of Indemnity in respect of any one item

2. the Property belongs to or is the responsibility of the Insured

3. the Property is

    (a) on or adjacent to the site of the Contract Works within the Territorial Limits or

    (b) in transit within the Territorial Limits by road (whether under its own power or otherwise) rail or inland waterway or

    (c) elsewhere within the Territorial Limits (but not Item 1 of the Property Insured) and stored in a locked premises or compound.

Professional Fees

The Company will in addition to the Limit of Indemnity indemnify the Insured for architects surveyors consulting engineers and other professional fees necessarily incurred in the repair reinstatement or replacement of Damage to the Property Insured to which the indemnity provided by this Section applies

Provided that

(a) such fees shall not exceed that authorised under the scales of the appropriate professional body or institute regulating such charges

(b) the Company shall not indemnify the Insured against any fees incurred by the Insured in preparing or contending any claim.

Debris Removal

The Limit of Indemnity provided in respect of Item 1 of the Property Insured shall extend to include the cost and expenses necessarily incurred by the Insured with the consent of the Company in

(a) removing and disposing of debris from or adjacent to the site of the Contract Works

(b) dismantling or demolishing

(c) shoring up or propping

(d) cleaning or clearing of drains mains services gullies manholes and the like within the site of the Contract Works

consequent upon Damage for which indemnity is provided by this Section

Provided that the Company shall not be liable in respect of seepage pollution or contamination of any Property not insured by this Section.

Off-site Storage

The indemnity provided in respect of Item 1 of the Property Insured extends to apply to materials or goods whilst not on the site of the Contract Works but intended for incorporation therein where the Insured is responsible under contract conditions provided that the value of such materials and goods has been included in an interim certificate and they are separately stored and identified as being designated for incorporation in the Contract Works.

Final Contract Price

In the event of an increase occurring to the original price the Limit of Indemnity in respect of Item 1 of the Property Insured shall be increased proportionally by an amount not exceeding 20%.

Tools Plant Equipment and Temporary Buildings

The Limit of Indemnity in respect of Items 2, 3 and 5 of the Property Insured is subject to average and if at the time of any Damage the total value of such Item of the Property Insured is of greater value than the Limit of Indemnity the Insured shall be considered as being his own insurer for the difference and shall bear a rateable share of the loss accordingly.

Section 3 – Contract Works

Speculative Housebuilding

The insurance in respect of Item 1 of the Property Insured shall notwithstanding Exception 4(b) for private dwelling houses flats and maisonettes constructed by the Insured for the purpose of sale continue for a period up to 180 days beyond the date of practical completion pending completion of sale.

Practical completion shall for the purposes of this extension mean when the erection and finishing of the private dwelling house are complete apart from any choice of decoration fixtures and fittings which are left to be at the option of the purchaser.

Local Authorities

The Indemnity provided in respect of Item 1 of the Property Insured shall include any additional cost of reinstatement consequent upon Damage to the Property Insured which is incurred solely because of the need to comply with building or other regulations made under statutory authorityor with bye-laws of any Municipal or Local Authority

Provided that

1. the Company shall not indemnify the Insured against the cost of complying with such regulations or bye-laws

   (a) in respect of Damage which is not insured by this Section

   (b) if notice has been served on the Insured by the appropriate authority prior to the occurrence of such Damage

   (c) in respect of any part of the Insured Property which is undamaged other than the foundations of that part which is the subject of Damage

2. the Company shall not indemnify the Insured against any rate tax duty development or other charge or assessment arising out of capital appreciation which may be payable in respect of the Property by its owner by reason of compliance with such regulations or bye-laws

3. reinstatement is commenced and carried out with reasonable despatch

4. nothing in this extension shall increase the liability of the Company to pay any amount exceeding the Limit of Indemnity in any one Period of Insurance.

Immobilised Plant

The indemnity provided in respect of Items 2 and 4 of the Property Insured shall include the cost of recovery or withdrawal of unintentionally immobilised constructional plant or equipment provided that such recovery is not necessitated solely by reason of electrical or mechanical breakdown or derangement.

Free Materials

Property for which the Insured is responsible shall include all free materials supplied by or on behalf of the Employer (named in the contract or agreement entered into by the Insured)

Provided that the total value of all such materials shall be included in the Limit of Indemnity for Item 1 of the Property Insured and also included in the declaration made to the Company under Condition 2.

Exceptions

The Company shall not indemnify the Insured against

1. the cost and expenses of replacing or making good any of the Property Insured which is in a defective condition due to faulty defective or incorrect

   (a) workmanship

   (b) design or specification

   (c) materials goods or other property installed erected or intended for incorporation in the Contract Works

   but this exception shall not apply to accidental Damage which occurs as a direct consequence to the remainder of the Property Insured which is free of such defective condition.

2. Damage due to

   (a) wear tear rust or other gradual deterioration

   (b) normal upkeep or normal making good

   (c) disappearance or shortage which is only revealed when an inventory is made or is not traceable to an identifiable event.

Section 3 – Contract Works

3. Damage to

  (a) machinery plant tools or equipment due to its own explosion breakdown or derangement but this exception shall be limited to that part responsible and shall not extend to other parts which sustain direct accidental Damage therefrom

  (b) aircraft hovercraft or watercraft other than hand propelled watercraft or other watercraft not exceeding 20ft in length

  (c) any mechanically propelled vehicle (including trailer attached thereto) whilst being used in circumstances for which compulsory insurance or security is required under any legislation governing the use of the vehicle but this exception shall not apply to

    i) plant capable of digging below its own wheel base

    ii) mobile cranes

    iii) dumpers primarily designed for use on site

    iv) site clearing and levelling plant primarily designed for use on site

    v) any vehicle (the primary purpose for which is not the conveyance of plant materials or goods by road) with plant permanently attached

  (d) bank notes cheques securities for money deeds or stamps

  (e) structures (or any fixtures fittings or contents thereof) existing at the time of commencement of the Contract Works

  (f) Item 1 of the Property Insured in respect of any contract or development

    i) the value or anticipated cost of which at the time of its commencement exceeds the Limit of Indemnity for Item 1

    ii) the period for which at the time of its commencement exceeds the Maximum Period.

4. Damage to the Contract Works or any part thereof

  (a) caused by or arising from use or occupancy other than for performance of the contract or for completion of the Contract Works by or on behalf of the Insured

  (b) occurring after practical completion or in respect of which a Certificate of Completion has been issued unless such Damage arises

    i) during any period (other than the Maintenance Period) not exceeding 14 days following practical completion or issue of such Certificate in which the Insured shall remain responsible under the terms of the contract for the Contract Works or the completed part thereof

    ii) during the Maintenance Period and from an event occurring prior to the commencement thereof

    iii) by the Insured in the course of any operations carried out in pursuance of any obligation under the contract during the Maintenance Period.

5. Damage for which the Insured is relieved of responsibility under the terms of any contract or agreement.

6. (a) liquidated damages or penalties for delay or non-completion

  (b) consequential loss of any nature.

7. Damage occasioned by pressure waves caused by aircraft or other aerial devices travelling at sonic or supersonic speeds.

8. the Excess specified in the Schedule.

9. Damage in Northern Ireland caused by or happening through or in consequence of

  (a) civil commotion

  (b) any unlawful wanton or malicious act committed maliciously by a person or persons acting on behalf of or in connection with any unlawful association

For the purpose of this exclusion

  i) unlawful association means any organisation which is engaged in terrorism and includes any organisation which at the relevant time is a proscribed organisation within the meaning of the Northern Ireland (Emergency Provisions) Act 1973

  ii) terrorism means the use of violence for political ends and includes any use of violence for the purpose of putting the public in fear

In any suit action or other proceedings where the Company alleges that by reason of this Exception any Damage is not covered by this Section the burden of proving that such Damage is covered shall be on the Insured.

Section 4 – 21.2.1

In the event of the Insured entering into any contract or agreement by which the Insured is required to effect insurance under the terms of Clause 21.2.1 of the Joint Contracts Tribunal Standard Form of Building Contract (or any subsequent revision or substitution thereof) or under the terms of any other contract requiring insurance of like kind the Company will indemnify the Employer in respect of any expense liability loss claim or proceedings which the Employer may incur or sustain by reason of Damage to any property other than the Contract Works occurring during the Period of Insurance within the Territorial Limits and caused by

(a) collapse

(b) subsidence

(c) heave

(d) vibration

(e) weakening or removal of support

(f) lowering of ground water

arising out of and in the course of or by reason of the carrying out of the Contract Works

Provided that

1. the Company shall not be liable for any amount exceeding the Limit of Indemnity

2. the Insured shall notify the Company within 21 days of entering into or commencing work under such contract or agreement whichever is the sooner together with full details of the contract

3. once notified the Company may give 14 days notice to cancel the cover granted by this Section in respect of such contract or agreement or alternatively provide a quotation which may vary the terms of this Section

4. the indemnity provided by this Section in respect of such contract or agreement shall terminate 14 days from the date of issue of the quotation if the quotation has not by then been accepted by the Insured or the Employer.

Employer

For the purpose of this Section Employer shall mean any person firm company ministry or authority named as the Employer in the contract or agreement entered into by the Insured.

Exceptions

The Company shall not indemnify the Employer

1. against any expense liability loss claim or proceedings

    (a) caused by the negligence omission or default of the Insured or any agent or Employee of the Insured or of any sub-contractor or his employees or agents

    (b) which is attributable to errors or omissions in the planning or the designing of the Contract Works

    (c) arising from Damage which could reasonably be foreseen to be inevitable having regard to the nature of the work to be executed or the manner of its execution

    (d) arising from Damage to property which is at the risk of the Employer under the terms of the contract or agreement

    (e) arising from Contractual Liability

    (f) arising from Damage occasioned by pressure waves caused by aircraft or other aerial devices travelling at sonic or supersonic speeds.

2. if the contract or agreement specifies that shoring of any building or structure is required and such shoring is necessary within 35 days of commencement of the contract or agreement.

3. against any expense liability loss claim or proceedings arising from

    (a) demolition or partial demolition of any building or structure

    (b) the use of explosives

    (c) tunnelling or piling work

    (d) underpinning

    (e) deliberate dewatering of the site.

4. in respect of any sum payable under any penalty clause or by reason of breach of contract.

5. against the Excess specified in the Schedule.

Extensions

Extensions to Sections 1 and 2 only

(a) Costs

The Company will in addition to the indemnity granted by each section pay

i) for all costs and expenses recoverable by any claimant from the Insured

ii) the solicitors fees incurred with the written consent of the Company for representation of the insured at

    (a) any coroner's inquest or fatal accident inquiry

    (b) proceedings in any Court arising out of any alleged breach of a statutory duty resulting in Bodily Injury or Damage to Property

iii) all costs and expenses incurred with the written consent of the Company in respect of a claim against the Insured to which the indemnity expressed in this Policy applies.

(b) Legal Defence

Irrespective of whether any person has sustained Bodily Injury the Company will at the request of the Insured also pay the costs and the expenses incurred in defending any director manager partner or Employee of the Insured in the event of such a person being prosecuted for an offence under the Health and Safety at Work etc. Act 1974 or the Health and Safety at Work (Northern Ireland) Order 1978.

The Company will also pay the costs incurred with its written consent in appealing against any judgement given

Provided that

(a) the offence was committed during the Period of Insurance

(b) the indemnity granted hereunder does not

    i) provide for the payment of fines or penalties

    ii) apply to prosecutions which arise out of any activity or risk excluded from this Policy

    iii) apply to prosecutions consequent upon any deliberate act or omission

    iv) apply to prosecutions which relate to the health safety or welfare of any Employee unless Section 1 is operative at the time when the offence was committed

    v) apply to prosecutions which relate to the health safety or welfare of any person not being an Employee unless Section 2 is operative at the time when the offence was committed

(c) the director manager partner or Employee shall be subject to the terms exceptions and conditions of the Policy in so far as they can apply.

(c) Indemnity to Other Persons

The Company will indemnify the following as if a separate Policy had been issued to each

(a) in the event of the death of the Insured the personal representatives of the Insured in respect of liability incurred by the Insured

(b) at the request of the Insured

    i) any officer or member of the Insured's canteen clubs sports athletic social or welfare organisations and first aid fire security and ambulance services in his respective capacity as such

    ii) any director partner or Employee of the Insured while acting in connection with the Business in respect of liability for which the Insured would be entitled to indemnity under this Policy if the claim for which indemnity is being sought had been made against the Insured

Provided that

(a) any persons specified above shall as though they were the Insured be subject to the terms exceptions and conditions of this Policy in so far as they can apply

(b) nothing in this extension shall increase the liability of the Company to pay any amount exceeding the Limit of Indemnity of the operative Section(s) regardless of the number of persons claiming to be indemnified.

Extension to Sections 1, 2 and 3 only

(d) Indemnity to Principal

Where any contract or agreement entered into by the Insured for the performance of work so requires the Company will

(a) indemnify the Principal in like manner to the Insured in respect of the principal's liability arising from the performance of the work by the Insured

(b) note the interest of the Principal in the Property Insured by Section 3 to the extent that the contract or agreement requires such interest to be noted.

Extension to Section 2 only

(e) Cross Liabilities

The Company will indemnify each insured to whom this Policy applies in the same manner and to the same extent as if a separate policy had been issued to each provided that the total amount of compensation payable shall not exceed the Limit of Indemnity regardless of the number of persons claiming to be indemnified

Provided that the Company shall not indemnify the Insured against liability for which an indemnity is or would be granted under any Employers Liability Insurance but for the existence of this Policy.

General Exceptions

The Company shall not indemnify the Insured

1. i)   for loss destruction of or damage to any property whatsoever or any loss or expense whatsoever resulting or arising therefrom or any consequential loss

    ii) for any legal liability of whatsoever nature

    directly or indirectly caused by or contributed to by or arising from

    (a) ionising radiations or contamination by radio-activity from any nuclear fuel or from any nuclear waste from the combustion of nuclear fuel

    (b) the radioactive toxic explosive or other hazardous properties of any explosive nuclear assembly or nuclear component thereof.

    In respect of Bodily Injury caused to an Employee this Exception shall apply only when the Insured under a contract or agreement has undertaken to indemnify a Principal or has assumed liability under contract for such Bodily Injury and which liability would not have attached in the absence of such contract or agreement.

2. under Sections 1 or 2 in respect of Contractual Liability unless the sole conduct and control of claims is vested in the Company but the Company will not in any event indemnify the Insured in respect of

    i)   liquidated damages or liability under any penalty clause

    ii) Damage to Property which comprises the Contract Works and occurs after the date referred to in Exception 3 of Section 2 if liability attaches solely by reason of the contract

    iii) Damage against which the Insured is required to effect insurance under the terms of Clause 21.2.1 of the Joint Contracts Tribunal Standard Form of Building Contract (or any subsequent revision or substitution thereof) or under the terms of any other contract requiring insurance of like kind.

3. under Sections 2, 3 or 4 for any consequence of war invasion act of foreign enemy hostilities (whether war be declared or not) civil war rebellion revolution insurrection or military or usurped power.

Conditions of the Policy

This policy and the Schedule shall be read together and any word or expression to which a specific meaning has been attached in any part of this Policy or of the Schedule shall bear such meaning wherever it may appear.

1. Alteration in Risk

The Company shall not be liable under this Policy if the risk be materially increased without the written consent of the Company.

2. Premium Adjustment

If the premium for this Policy is based on estimates an accurate record containing all particulars relative thereto shall be kept by the Insured

The Insured shall at all times allow the Company to inspect such records and shall supply such particulars and information as the Company may require within one month from the expiry of each Period of Insurance and the premium shall thereupon be adjusted by the Company (subject to the Minimum Premium chargeable for the risk being retained by the Company).

3. Duties of The Insured

The Insured shall take all reasonable care

(a)   to prevent any event which may give rise to a claim under this Policy

(b)   to maintain the premises plant and everything used in the Business in proper repair

(c)   in the selection and supervision of Employees

(d)   to comply with all statutory and other obligations and regulations imposed by any authority.

4. Make Good Defects

The Insured shall make good or remedy any defect or danger which becomes apparent and take such additional precautions as circumstances may require.

5. Maximum Payments

The Company may at any time at its sole discretion pay to the Insured the Limit of Indemnity (less any sum or sums already paid in respect or in lieu of damages) or any lesser sum for which

the claim or claims against the Insured can be settled and the Company shall not be under any further liability in respect of such claim or claims except for costs and expenses incurred prior to such payment

Provided that in the event of a claim or series of claims resulting in the liability of the Insured to pay a sum in excess of the Limit of Indemnity the Company's liability for costs and expenses shall not exceed an amount being in the same proportion as the Company's payment to the Insured bears to the total payment made by or on behalf of the Insured in settlement of the claim or claims.

6. Claims

The Insured or his legal personal representatives shall give notice in writing to the Company as soon as possible after any event which may give rise to liability under this Policy with full particulars of such event. Every claim notice letter writ or process or other document served on the Insured shall be forwarded to the Company immediately on receipt. Notice in writing shall also be given immediately to the Company by the Insured of impending prosecution inquest or fatal inquiry in connection with any such event. No admission offer promise payment or indemnity shall be made or given by or on behalf of the Insured without the written consent of the Company. In the event of Damage by theft or malicious act the Insured shall also give immediate notice to the police.

7. Subrogation

The Company shall be entitled if it so desires to take over and conduct in the name of the Insured the defence or settlement of any claim or to prosecute in the name of the Insured for its own benefit any claim for indemnity or damages or otherwise and shall have full discretion in the conduct of any proceedings and in the settlement of any claim and the Insured shall give all such information and assistance as the Company may require.

8. Contribution

If at the time of any event to which this Policy applies there is or but for the existence of this Policy there would be any other insurance covering the same liability or Damage the Company shall not be liable under this Policy except in respect of any excess beyond the amount which would be payable under such other insurance had this Policy not been effected.

Conditions of the Policy

9. Cancellation

The Company may cancel this Policy by giving thirty days notice by recorded delivery letter to the last known address of the Insured. The Company shall make a return of the proportionate part of the premium in respect of the unexpired Period of Insurance or if the premium has been based wholly or partly upon estimates the premium shall be adjusted in accordance with Condition 2.

10. Disputes

Any dispute concerning the interpretation of the terms of this Policy shall be resolved in accordance with the jurisdiction of the territory in which this Policy is issued.

11. Rights

In the event of Damage for which a claim is or may be made under Section 3

(a) the Company shall be entitled without incurring any liability under this Policy to

    i) enter any site or premises where Damage has occurred and take and keep possession of the Property Insured

    ii) deal with any salvage as they deem fit

but no property may be abandoned to the Company

(b) if the Company elects or becomes bound to reinstate or replace any property the Insured shall at their own expense produce and give to the Company all such plans and documents books and information as the Company may reasonably require. The Company shall not be bound to reinstate exactly or completely but only as circumstances permit and in reasonably sufficient manner and shall not in any case be bound to expend in respect of any one of the items of Property Insured more than the Limit of Indemnity in respect of such item.

12. Observance

The due observance and fulfilment of the terms exceptions conditions and endorsements of this Policy in so far as they relate to anything to be done or complied with by the Insured and the truth of the statements and answers in the proposal shall be conditions precedent to the liability of the Company to make any payment under this Policy.

Contractors' Insurance

Schedule

Policy Number :

Insured :   SPECIMEN POLICY
CONTRACTORS ALL RISKS

Address :   MARSLAND HOUSE
MARSLAND ROAD
SALE, CHESHIRE
M33 3AQ

Business :   BUSINESS DESCRIPTION OF THE INSURED AS RELEVANT TO THIS
POLICY OF INSURANCE

Period of Insurance : From
                        To
RENEWAL DATE   :

FIRST PREMIUM   :   £

ANNUAL PREMIUM   :   £

MINIMUM PREMIUM (See Condition 2)
SECTION 1 : £
SECTION 2 : £
SECTION 3 : £
SECTION 4 : £

MAXIMUM PERIOD (for the purposes of Section 3)
12 MONTHS PLUS 12 MONTHS MAINTENANCE PERIOD

Endorsements Applicable
NONE

Effective Date :

Reason for Issue : NEW BUSINESS

P.F. :

Contractors' Insurance

Policy Number :

Effective Date :

Schedule (cont'd)

Limits of Indemnity :

Section 1   Employers Liability :

Section 2   Public Liability : £
                    THIS LIMIT APPLIES IN RESPECT OF ANY ONE
                    OCCURRENCE OR SERIES OF OCCURRENCES ARISING OUT OF
                    ANY ONE CAUSE

Section 3   Contract Works :
Property Insured   **Limit of Indemnity**

Item 1 – Contract Works   £

Item 2 – Constructional Plant Tools and Equipment   £
              owned by the Insured

Item 3 – Temporary Buildings and Site Huts   £
              (including fixtures and fittings therein)

Item 4 – Hired–in or borrowed Property
              described in Items 2 and 3 not exceeding   £
                                                              any one item

Item 5 – Personal Effects and Tools of the Insureds
              Employees not exceeding £   any one Employee   £

Section 4   21. 2. 1. : £
                    THIS LIMIT APPLIES IN RESPECT OF ANY ONE
                    OCCURRENCE OR SERIES OF OCCURRENCES ARISING OUT OF
                    ONE CAUSE

Excess :
Section 2 :
        IN RESPECT OF DAMAGE TO PROPERTY ARISING FROM THE
        USE OF HEAT AWAY FROM THE INSURED'S PREMISES   £
        IN RESPECT OF DAMAGE TO PROPERTY FROM ANY OTHER
        CAUSE   £

Section 3 : DAMAGE CAUSED BY THEFT OR MALICIOUS DAMAGE £
        DAMAGE TO PERSONAL EFFECTS AND TOOLS OF THE
        INSURED'S EMPLOYEES £

Section 4 : £   IN RESPECT OF ALL CLAIMS IN THE AGGREGATE FOR ANY
        ONE CONTRACT OR AGREEMENT

Appendix 3

ABI model form of guarantee bond and House of Lords decision in *Trafalgar House* case

Part 1: ABI notes to the consultation document 1994

ABI CONSTRUCTION ON DRAFT
MODEL FORM OF GUARANTEE BOND FOR USE IN
CONSTRUCTION CONTRACTS

1. In the light of growing concern by the construction industry regarding the increasing use by Employers of "on demand" bonds for construction contracts and the statement by the Secretary of State for the Environment that "on demand bonds have no place in government contracts", the Surety Bond Panel of the Association of British Insurers met in late 1993 to consider the subject of a Model Form of Guarantee Bond.

2. These discussions were given added impetus by the publication of the 'Latham Report' in July 1994 and the decisions of the Court of Appeal in England in early 1994 in the cases of Trafalgar House (now listed for Appeal in the House of Lords) and Perar BV.

3. The Association has now produced a draft Model Form of Guarantee Bond which seeks to avoid the use of archaic language and to express the obligations of the parties clearly. The draft Model Form of Guarantee Bond now distributed for discussion makes it clear that the obligations of the Surety relate to the underlying contract and that the Surety is liable to the extent that the Contractor itself is liable for any breach of contract, subject to the bond amount.

4. The draft Model Form of Guarantee Bond is designed to protect Employers against the consequences of default (usually insolvency) by the Contractor. If there has been a breach of contract by the Contractor and the Employer is entitled to recover any sums under the contract or damages, then a claim can be made on the Surety for the recovery of such sums or damages due from the Contractor up to the bond amount.

5. The draft Model Form of Guarantee Bond strikes a fair balance between Contractor, Employer and Surety and should be distinguished from an "on demand" bond.

6. Insolvency alone does not entitle the Employer to payment unless, as a consequence, losses and damages are payable to the Employer by the Contractor (which was the effect of the decision in Perar). Most construction contracts provide for sums to become due and payable to the Employer following any

default (including insolvency) and the draft Model Form of Guarantee Bond safeguards the payment of those sums.

7. Employers accepting Guarantee Bonds in the draft Model Form will be entitled, subject to the bond amount, to recover all sums which fall due to them in accordance with the conditions of the contract if those sums are not paid by the Contractor.

8. The Association invites comments upon the draft Model Form of Guarantee Bond from the construction industry and all other interested parties. These should be sent to Mr T Humphreys, Association British Insurers, 51 Gresham Street, London EC2V 7HQ before 30 December 1994.

Part 2: The Model Form of Guarantee Bond (September 1995)

THIS GUARANTEE BOND is made as a deed BETWEEN the following parties whose names and [registered office] addresses are set out in the Schedule to this Bond (the "Schedule")

1. The "Contractor" as principal
2. The "Guarantor" as guarantor, and
3. The "Employer"

WHEREAS

1. By a contract (the "Contract") entered into or to be entered into between the Employer and the Contractor particulars of which are set out in the Schedule the Contractor has agreed with the Employer to execute works ("the Works") upon and subject to the terms and conditions therein set out

2. The Guarantor has agreed with the Employer at the request of the Contractor to guarantee the performance of the obligations of the Contractor under the Contract upon the terms and conditions of this Guarantee Bond subject to the limitation set out in clause 2

NOW THIS DEED WITNESSES as follows:

1. The Guarantor guarantees to the Employer that in the event of a breach of the Contract by the Contractor the Guarantor shall subject to the provisions of this Guarantee Bond satisfy and discharge the damages sustained by the Employer as established and ascertained pursuant to and in accordance with the provisions of or by reference to the Contract and taking into account all sums due or to become due to the Contractor

2. The maximum aggregate liability of the Guarantor and the Contractor under this Guarantee Bond shall not exceed the sum set out in the Schedule (the "Bond Amount") but subject to such limitation and to clause 4 the liability of the Guarantor shall be co-extensive with the liability of the Contractor under the contract

3. The Guarantor shall not be discharged or released by any alteration of any of the terms conditions and provisions of the Contract or in the extent or nature of the Works and no allowance of time by the Employer under or in respect of the Contract or the Works on the part of the Employer shall in any way release reduce or affect the liability of the Guarantor under this Guarantee Bond

4. Whether or not this Guarantee Bond shall be returned to the Guarantor the obligations of the Guarantor under this Guarantee Bond shall be released and discharged absolutely upon Expiry (as defined in the Schedule) save in respect of any breach of the Contract which has occurred and in respect of which a claim in writing containing particulars of such breach has been made upon the Guarantor before Expiry

5. The Contractor having requested the execution of this Guarantee Bond by the Guarantor undertakes to the Guarantor (without limitation of any other rights and remedies of the Employer or the Guarantor against the Contractor) to perform and discharge the obligations on its part set out in the Contract

6. The Guarantee Bond and the benefits thereof shall not be assigned without the prior written consent of the Guarantor and the Contractor

7. This Guarantee Bond shall be governed by and construed in accordance with the laws of [England and Wales] [Scotland] and only the courts of [England and Wales] [Scotland] shall have jurisdiction hereunder

THE SCHEDULE

The Contractor: [   ] whose *[address]* registered
                     office address is at [   ]

The Guarantor: [   ] whose registered
                     office address is at [   ]
                     [   ]

The Employer: [   ] whose *[address]* registered
                     office address is at [   ]

The Contract: A contract [dated the   day of   ] *[to be entered into]*
                     between the Employer and the Contractor in the form known
                     as [   ] for the construction of works comprising
                     [   ] for the original contract sum of
                     [   ] pounds (£[   ])

The Bond Amount: The sum of £[   ] pounds sterling (£[   ])
                     [Insert any provisions for reduction of the Bond Amount]

Expiry: *[Insert details of the event agreed between the parties]* which shall be
                     conclusive for the purposes of this Guarantee Bond

IN WITNESS whereof the Contractor and the Guarantor have executed and delivered this Guarantee Bond as a Deed this   day of   One Thousand Nine Hundred and

EXECUTED AND DELIVERED as a deed by
CONTRACTOR

EXECUTED AND DELIVERED as a deed by
GUARANTOR

Part 3: *Insurance Industry view of the* Trafalgar House *case*

PERFORMANCE BONDS LANDMARK CASE
HOUSE OF LORDS JUNE 1995
GENERAL SURETY & GUARANTEE CO LIMITED
V.
TRAFALGAR HOUSE CONSTRUCTION (REGIONS) LTD

1. (i)   In June this year the House of Lords reversed the Court of Appeal Judgment
          of February 1994 in the case of *Trafalgar House Construction (Regions) Limited
          v. General Surety & Guarantee Co Limited.*
    (ii) The Court of Appeal had previously decided that the ICE Form of Bond was
          not a Guarantee, was payable on demand and that no credit should be given
          for any unpaid sums due to the Contractor in reduction of any claim.

2. The House of Lords unanimously found that:
    (i)   in line with over 150 years of established practice, Trafalgar's ICE type bond
          issued by GSG operated as a guarantee of the obligations of the Contractor
          under the contract and of any debt due to the Employer under the contract
          following the insolvency of the Contractor.
    (ii) the Bond secure payment of the Employer's damages upon Contractor
          default.
    (iii) in calculating damages, contractual balances had to be brought into account
          and that the Surety would be liable to the same extent as the Contractor and
          entitled to rely upon the same defences.
    (iv) archaic and lengthy bond wordings should be avoided.

In his judgment, Lord Jauncey also commented, 'In recent years there has come
into existence a creature described as an "on demand bond" in terms of which
the creditor is entitled to be paid merely on making a demand for the amount of
the bond. . . . All that was required to activate it was a demand by the creditor
stated to be on the basis of the event specified in the bond.'

3. The Association of British Insurers had been awaiting the outcome of the House
    of Lords appeal before publishing a new Model Form of Guarantee Bond which it
    hopes will become an industry standard in the future.

This model Form of Guarantee Bond will shortly be promoted by the ABI with
the support of many contractors. The NJCC have also indicated that they would
like to adopt this Model Form for general use. Hopefully these reins will also be
taken up by the legal profession generally, the Royal Institute of Chartered
Surveyors, the Building Employers Confederation and, without more ado, the
Institute of Civil Engineers. In the absence of any lead from the Government
including the well overdue Statute requiring all publicly let jobs to be adequately
protected by Performance Bonds, at least private employers and their representatives in the various professions will be better prepared with the knowledge
of the commercial purpose of such a bond which will save the industry considerable time and money. Additionally, one would trust that in the near future,
all Government Departments, Local Authorities and Public Utilities will take the
lead in adopting the common form which is certainly the case in the United States
and in a number of other European Countries now using the International
Chamber of Commerce bond wording.

To this end we in the insurance market should endorse the new ABI Model Form of Guarantee and with everyone's support, hope it will become an industry standard in the very near future.

Appendix 4

Specimen form of general counter indemnity

THIS INDEMNITY is made BETWEEN the Undersigned Companies named in the Schedule of Indemnitors (hereinafter called "the Indemnitors") of the one part and (hereinafter called "the Surety") of the other part

WHEREAS the Surety has executed or procured or may hereafter execute or procure another or others to execute a Bond or Bonds for securing the obligations of any one or more of the Principals named in the Schedule of Principals hereto (such Bond or Bonds being hereinafter defined)

NOW THIS DEED WITNESSETH and it is hereby agreed and declared as follows:

1. The Indemnitors shall indemnify and keep indemnified the Surety against claims liabilities costs expenses and losses (including interest) incurred by the Surety under or by virtue of any Bond or Bonds. Costs and expenses shall include those incurred by the Surety where a claim or threatened claim is resisted or disposed of without payment being made in respect of such claim. The obligations of the Indemnitors under this Clause are not to be reduced or qualified in any way by the following provisions of this Deed.

2. In the event of any payment being made by the Surety in respect of any claim liability costs expenses and/or losses under or by virtue of any Bond or Bonds the Indemnitors shall repay the full amount thereof to the Surety upon written request by the Surety and it is expressly agreed that the acknowledgement receipt or other voucher in respect of any such payment made by the Surety shall as between the Indemnitors and the Surety be conclusive evidence that the amount of such payment has been made and is accordingly repayable hereunder by the Indemnitors to the Surety.

3. In the event of the liquidation of any one or more of the Indemnitors or Principals (except a voluntary liquidation for the purpose of reconstruction or amalgamation) or on the appointment of a Receiver in respect of any of the assets of or the undertaking of any such Indemnitor or Principal then the Indemnitors will if requested in writing by the Surety deposit with the Surety in cash the full amount of any Bond or Bonds then outstanding on behalf of any such Principal or Principals if more than one. Such monies shall be paid by the Indemnitors to the Surety within fourteen days after such request and the Surety may use such monies to settle or pay any claim or claims together with any costs expenses and/or losses incurred by the surety under or by virtue of any such Bond or Bonds. If there should be a surplus in the hands of the Surety after paying or settling all such claims costs expenses and/or

losses such surplus shall be refunded to the Indemnitors together with an annual interest thereon calculated at the Base Rate of Lloyds Bank Plc from time to time in force.

4. The Surety shall not be obliged to obtain the consent of or to notify the Indemnitors before or after the execution of any such Bond or Bonds (including any extension or renewal thereof). Furthermore the rights and remedies of the Surety under this Deed shall be additional to and not in reduction or in lieu of any other rights and remedies which the Surety may already have or may hereafter acquire against the Indemnitors or any other persons whether by the terms of any other instrument or by operation of law.

5. In addition to the obligations of the Indemnitors hereinbefore contained the Indemnitors shall procure the payment of or pay on request all premiums taxes and any other expenses payable by any Principal or Principals to the Surety in respect of any Bond or Bonds.

6. The Indemnitors hereby expressly agree that all obligations and liabilities of the Indemnitors hereunder are joint and several and that the Surety may enforce any or all of its rights and remedies hereunder against any persons liable and in any order or priority as the Surety may in its absolute discretion think fit.

7. If any of the companies named in the Schedule hereto as Indemnitors shall fail to execute this Deed or if the execution hereof by any such Indemnitor shall be invalid or unenforceable for any reason whatsoever the provisions hereof shall nevertheless be binding upon and continue in full force and effect as regards any Indemnitor who has duly executed the same.

8. If the Surety shall procure the execution of any Bond or Bonds by another as a Procured Sole Surety or by another as a Procured Co-Surety then this Deed shall also operate for the benefit of such Procured Sole Surety or Co-Surety and the Surety is irrevocably authorised to extend the benefit of this Deed to any such Procured Sole Surety or Co-Surety so far as such extension may be required to protect the interests of such Procured Sole Surety or Co-Surety without reducing or otherwise qualifying the obligations of the Indemnitors to the Surety hereunder.

9. Bond or Bonds means any Bond or Guarantee Indemnity or other obligatory instrument in whatever form executed by or procured by the Surety for securing the obligations of any one or more of the Principals and shall include any reinsurance accepted by the Surety and any indemnity or guarantee given by the Surety in respect of any Bond or Bonds procured as aforesaid. This Indemnity shall operate for the benefit of the Surety in respect of any Bond or Bonds executed jointly with another or others and shall apply to any Bond or Bonds already executed or procured as those hereafter executed or procured.

10. Any communications hereunder may be made by sending the same in a pre-paid letter by Registered Post or Recorded Delivery Service addressed to any party hereto at the last known address or Registered Office of such party as the case may be.

11. The Indemnitors hereby confirm by executing this Deed that each of them is lawfully empowered by its Memorandum of Association or otherwise to enter into this Deed and that each is duly authorised to execute the same in the manner appearing below.

12. The law governing this Deed shall be English law and only the Courts of England and Wales shall have jurisdiction hereunder.

SCHEDULE OF PRINCIPALS

SCHEDULE OF INDEMNITORS

IN WITNESS WHEREOF the Indemnitors have hereunto affixed their Common Seals this   day of   199 .

Appendix 5

Outline of Part I of the Unfair Contract Terms Act 1977

A short summary of the purposes of each section of Part I is given for easy reference.

Unfair Contract Terms Act 1977 (Part I)

Avoidance of Liability

Section 1
Defines Negligence
Sub-section
(1) Breach of
    (a) a contractual term to take reasonable care;
    (b) any common law duty to take reasonable care;
    (c) the common duty of care imposed by the Occupiers' Liability Act, 1957.

Section 2
For Negligence
Sub-section
(1) for death or personal injury – invalid;
(2) for other loss or damage – subject to test of reasonableness;
(3) awareness of or agreement to contract term or notice is not a voluntary acceptance of any risk.

Section 3
For Breach of Contract
Sub-section
(1) either where one deals as consumer *or* on the other's written standard terms of business;
(2) subject to the test of reasonableness.

Section 4
By Indemnity Clauses
Sub-section
(1) a person dealing as consumer cannot be made to indemnify another for his negligence or breach of contract except if reasonable;
(2) whether the liability is
    (a) direct or vicarious;
    (b) to the consumer or someone else.

Section 12
Defines 'Deals as a Consumer'
Sub-section
(1) (a) he does not make the contract in the course of a business;
    (b) the other party does make the contract in this way;
    (c) in the case of sections 6 and 7 the goods are ordinarily supplied for private use or consumption.

Liability arising from Sale or Supply of Goods

Section 5
'Guarantee' of Consumer Goods
Sub-section
(1) no guarantee can exclude or restrict liability for loss or damage to goods supplied which results from:
    (a) defects while in consumer use;
    and
    (b) negligence of a manufacturer or distributor.
(2) defines
    (a) 'in consumer use'; and
    (b) a guarantee.
(3) excludes guarantees between parties to a contract under which possession or ownership of the goods passed.

Section 6
Sale and Hire Purchase
Sub-section
(1) liability for breach of obligations arising from
    (a) section 12 of the Sale of Goods Act 1893;
    (b) section 8 of the Supply of Goods (Implied Terms) Act 1973
cannot be excluded or restricted by any contract term.
(2) liability against a person dealing as consumer from:
    (a) sections 13, 14, 15 of the 1893 Act;
    (b) sections 9, 10, 11 of the 1973 Act;
cannot be excluded or restricted by any contract term.
(3) liability in (2) against a non-consumer is subject to the requirements of reasonableness.

Section 7
Miscellaneous Contracts under which Goods Pass
Sub-section
(1) contracts other than sale of goods or hire purchase with contract terms excluding or restricting liability for breach of obligation arising by implication of law;
(2) as against a consumer (in respect of correspondence of goods with description or sample, or quality or fitness for purpose) are invalid;
(3) as against a non-consumer are subject to a requirement of reasonableness.

Other Provisions

Section 8
Misrepresentation
Sub-section
(1) amends section 3 of the Misrepresentation Act 1967

Section 9
Effect of Breach
Sub-section
(1) amends the law concerning fundamental breach.

Section 10
Evasion by means of Secondary Contract
Not possible.

Section 11
The 'Reasonableness' Test
Sub-section
(1) the time to consider this in the case of a contract term is when the contract was made;
(2) in the case of sections 6 and 7 consider the guidelines in Schedule 2;
(3) the time to consider this in the case of a notice (not having contractual effect) is when the liability arose or (but for the notice) would have arisen;
(4) where a contract term or notice restricts liability to a specified sum, consider
  (a) the resources available to meet the liability; and
  (b) the availability of insurance.
(5) it is for those claiming a contract term or notice satisfies the 'reasonableness' test to show that it does.

Section 13
Varieties of Exemption Clause to which Part I applies

Section 14
Definitions of Part I
'Business' includes a profession and the activities of any government department or local or public authority. "Notice' includes an announcement, whether or not in writing and any other communication or pretended communication. 'Personal injury' includes any disease and any impairment of physical or mental condition.

Summary of Practice Note 22 Model Clauses

Model Clauses
for use where the Employer in using the Standard Form of Building Contract 1980 Edition incorporating Amendment 2: November 1986 does not wish to use either clause 22A or clause 22B or clause 22C.2 to .4 in respect of loss or damage to the Works: or clause 22C.1 in respect of loss or damage to the existing structures and contents

Model Clauses: New Building Work

22E: for use where the Employer does not wish to use 1986 clause 22A or 22B but wishes to take the *risk* of loss or damage to the Works from the risks covered by the 1986 definition "All Risks Insurance" but does **not** wish to insure.

8E: for Nominated Sub-Contracts NSC/4 and NSC/4a for use only where model clause 22E is included in the Main Contract.

22F: for use where the employer does not wish to use 1986 clause 22A or 22B but wishes to take the *sole risk* of loss or damage to the Works from the risks covered by the 1986 definition "All Risks Insurance" but does **not** wish to insure.

8F: for Nominated Sub-Contracts NSC/4 and NSC/4a for use only where model clause 22F is included in the Main Contract.

Model Clauses: Work in or extensions to existing buildings

22G: for use where the Employer does not wish to use 1986 clause 22C.2 to .4 but wishes to take the *risk* of loss or damage to the Works from the risks covered by the 1986 definition "All Risks Insurance" but does **not** wish to insure.

8G: for Nominated Sub-Contracts NSC/4 and NSC/4a for use only where model clause 22G is included in the Main Contract.

22H: for use where the Employer does not wish to use 1986 clause 22C.2 to .4 but wishes to take the *sole risk* of loss or damage to the Works from the risks covered by the 1986 definition "All Risks Insurance" but does **not** wish to insure.

8H: for Nominated Sub-Contracts NSC/4 and NSC/4a for use only where model clause 22H is included in the Main Contract.

22J: for use where the Employer does not wish to use 1986 clause 22C.1 but wishes to take the *risk* of loss or damage by the 1986 Specified Perils to the existing structures and the contents owned by the Employer or for which he is responsible but does **not** insure.

8J: for Nominated Sub-Contracts NSC/4 and NSC/4a for use only where model clause 22J is included in the Main Contract.

22K: for use where the Employer does not wish to use 1986 clause 22C.1 but wishes to take the *sole risk* or loss or damage by the 1986 Specified Perils to the existing structures and the contents owned by the Employer or for which he is responsible but does **not** wish to insure.

8K: for Nominated Sub-Contracts NSC/4 and NSC/4a for use only where model clause 22K is included in the Main Contract.

Model clauses 22E, 22G and 22J have been drafted on the basis that the term **"risk"** means in relation to the Contractor that the Employer is responsible for loss or damage by the risks covered by the 1986 definition "All Risks Insurance" but can have recourse against the Contractor if the loss or damage to the Works and Site Materials (22E and 22G)or to the existing structures and their contents owned by the Employer or for which he is responsible (22J) is caused by the negligence of the Contractor or of any Nominated or Domestic Sub-Contractor.

Model clauses 22F, 22H and 22K have been drafted on the basis that the term **"sole risk"** means in relation to the Contractor that the Employer is responsible for loss or damage by the risks covered by the 1986 definition "All Risks Insurance" and **cannot** have recourse against the Contractor **even if** the loss or damage to the Works and Site Materials (22F and 22H) or to the existing structures and their contents owned by the Employer or for which he is responsible (22K) is caused by the negligence of the Contractor or of any Nominated or Domestic Sub-Contractor.

It **is essential** to include the above meanings as definitions in the Contract Documents so as to avoid any doubt as to the meaning of the terms used.

Employers may wish to reduce or extend their risk for loss or damage to the Works and Site Materials or to the existing structures and their contents owned by the Employer or for which he is responsible from that given above. If so they should alter the relevant model clause accordingly, clearly define the meanings of the terms used and, as applicable, alter the consequential amendments to both the main contract and the nominated sub-contract.

In considering the model clauses 22E to 22K it must be remembered that they are based on the meanings given to "risk" (22E, 22G and 22J) and "sole risk" (22F, 22H and 22K) given above and interpreted accordingly.

If these model clauses or any adaptations thereof are used it would be advisable for the Employer to take **legal advice** thereon.

Eurotunnel Channel Fixed Link Project construction policy

The Thames Barrier insurance mentioned in Chapter 8 was a policy arranged by the main contractors for the civil engineering works lying at the heart of the project. That policy was complemented by separate construction insurances on an equally wide basis for the installation of gates and other equipment. The Eurotunnel Project insurance however goes one stage further in covering, under one single policy, all the works, equipment and rolling stock that make up the complete Eurotunnel Project both in France and in the United Kingdom.

The more parties involved and the more complex the project, the more sensible it becomes for construction insurance to be arranged on a comprehensive project basis. Eurotunnel's policy is a good example of this concept used sensibly and correctly. Its principal claim to fame is that it may well be the first truly Franco-British single contract of insurance.

Comprehensive project (sometimes referred to as wrap-up or omnibus) insurance usually covers at least public liability and all risks insurance of the works for all parties involved, or as many as possible. This avoids the time and trouble spent by all parties involved in any loss, damage or liability claims blaming the other parties to the contract or sub-contract. See the other advantages and disadvantages of project insurance given in Chapter 8.

The product that the construction industry wants from insurers is one policy covering the project and protecting all the parties involved in its construction, not separate policies covering the liability of its producers. This is difficult to achieve because of the structure of the insurance market, the risks involved and the general unwillingness of insurers to provide latent defects insurance where there is a lack of claims data or other statistics on building defects on which to base an appraisal of risk, and, in a small market, selection against the insurer. But, the report produced by the Insurance Feasibility Steering Committee through the National Economic Development Office (NEDO) entitled *Building User's Insurance Against Latent Defects (BUILD)*, is seen by some as the answer to the problem. However, this report only applies to commercial buildings and not to civil engineering projects. Probably the main obstacle to a full project insurance is the small size of, and bad experience suffered by, the professional indemnity insurance market in the construction industry.

The policy

The first thing that strikes the reader of this policy is its bilingual form and that the co-insurers subscribing directly represent a broad cross-section of both British and

French companies as well as Lloyd's Underwriters. The insured parties are listed separately depending upon their French or British identity, but taken together the comprehensive list includes:

(a) all companies in the Eurotunnel Group as principals (i.e. project owners) and any eventual successors;
(b) the banks and the other parties providing the finance to Eurotunnel under the latter's credit agreement;
(c) the Governments of United Kingdom and France and their appointed representatives;
(d) the consultant companies acting in the supervisory role of *maître d'oeuvre*;
(e) the ten British and French construction companies who in joint venture collectively constitute Transmanche Link as main contractor together with the parent and subsidiary companies of the individual firms;
(f) any consultants, sub-contractors of any tier or suppliers engaged by the contractors as well as any Government departments or authorities providing services to the project;
(g) any other person or firm engaged by the principal.

CAR cover was granted for the nearly six year period of construction up to mid-1993 with extensions of period held covered, followed by a maintenance period not exceeding 24 months.

Cover was in respect of the whole project defined as 'the design, procurement, construction, testing, commissioning and maintenance of the works for the Fixed Link between the United Kingdom and France'.

The insured property is defined under two items as:

1. Works comprising permanent works and temporary works ... including unfixed materials, goods and all other property for incorporation therein (which encompasses the shuttle trains, locomotives and spares ordered for Eurotunnel).
2. All plant and equipment and temporary buildings together with their contents and all other property owned by or for which the insured accepts responsibility as well as tools, equipment, clothing and personal effects of employees of the insured.

The combined sum insured on a first loss basis is £500 million, any one occurrence.

(At this point, it is worth noting that the same policy also contains a public liability section with a primary limit of £25 million, any one occurrence, as well as a third section covering additional interest charges payable in the event of prolonged delay directly caused by damage to works or plant insured for CAR risks.)

Coverage applies equally to all construction sites in France or England including transit between sites. Subject to certain restrictions regarding the manufacture and delivery of external supplies, the territorial limits are world-wide, excluding USA and Canada, and provision is also made for a modest amount of cover in respect of marine and/or air sendings if required subject to declaration.

The operative clause reads 'The Insurers will indemnify the Insured against physical loss of or damage to the Insured property howsoever caused subject to the following Exclusions ...'.

The exclusions themselves are those normally found in a CAR policy but many with qualifications negotiated to reflect the needs of a project policy covering both principal main contractors and sub-contractors on an equal footing.

(a) Aircraft are excluded but only boats in excess of 10 m.

(b) Insurers are not liable for any item of constructional plant or equipment due to its own mechanical or electrical breakdown, failure or derangement but this exclusion does not apply to specified types of resultant damage and does not apply to the machinery which is installed as part of the works.

(c) The normal defects exclusion is re-worded to exclude 'the costs necessary to replace, repair or rectify any defect in design, plan, specification, materials or workmanship but should unintended damage result from such defect, this exclusion shall be limited to the additional costs of improvement to the original design, plan or specification including the costs of carrying out such improvements'.

(d) On the other hand, an increased deductible of £250,000 applies to each occurrence, or series of occurrences, arising out of one event consequent upon defective design, materials or workmanship insofar as concerns damage to the works.

Insurers exclude:

costs and expenses in respect of tunnel and shaft excavations relating to:

.1 dewatering except where such cost results from an incident which suddenly produces water flows exceeding those which could reasonably be anticipated;
.2 overbreak excavation in excess of the maximum excavation provided for in the plans and the additional expenses resulting therefrom for refilling of cavities;
.3 loss destruction or damage in advance of the tunnel face.

There is also a deductible of £100,000 each occurrence in respect of off-shore activities, tunnelling and shaft operations.

In respect of all other loss or damage affecting the works (temporary works or materials), the deductible is £25,000 each occurrence.

In respect of constructional plant and equipment, the deductibles are:

(a) £100,000 each occurrence in respect of tunnelling machinery whilst in the running or service tunnels or whilst being placed below ground or whilst being removed therefrom;
(b) £2,500 in respect of all other loss or damage.

French law requires that separate decennale insurance be effected in respect of buildings (as opposed to civil engineering works) on French soil and the policy therefore excludes such damage occurring after completion which is compulsorily insurable.

Full all risks cover continued on the whole of the permanent works until the (single) completion certificate was issued to the contractors or until the works were put into commercial operation, whichever first occurred.

During the subsequent maintenance period of 12 months (or 24 months in respect of electrical and mechanical equipment) cover was granted in respect of physical loss or damage to:

(a) any outstanding work;
(b) the permanent works arising out of maintenance operations;

(c) the permanent works occurring during the maintenance period and arising from a cause happening prior to commencement of such period.

This latter broad coverage is however restricted to the extent that it does not benefit manufacturers of the rolling stock in respect of damage caused by any defect in design plan, specification, materials or workmanship in such equipment.

The CAR section of the combined project policy also contains explanatory memoranda dealing *inter alia* with:

(a) professional fees;
(b) debris removal and loss minimisation expenses;
(c) marine/non-marine loss sharing;
(d) the previously mentioned facility for transit by sea or air anywhere in Europe and general average;
(e) hired-in constructional plant and liability for continuing hire charges;
(f) expediting expenses not exceeding 50% of 'normal' indemnity;
(g) 72 hour clause in respect of the application of deductibles to storm, flood or earthquake.

The CAR coverage must, of course, be read in the context of the detailed clauses and also in the context of the general memoranda and general conditions applicable to a policy issued in three sections. It is not possible to reproduce these in full, particularly since they are repeated in both English and French.

This bilingual aspect, as well as the dual nationality of the policy itself is underlined by the general memorandum which states that the construction, validity and performance of the insurance shall be governed by and interpreted in accordance with French law to the extent that the claimant is French (or with English law to the extent that the claimant is English). In either case, as appropriate, it is the French or English version of the wording which is deemed to be authentic, although in the event of dispute as to interpretation, meanings and definitions in both languages may be taken into consideration to assist the resolution of any dispute.

To the extent that French law applies, the provisions of the French Code des Assurances are incorporated into the policy which also provides for arbitration on disputes as to the amount of indemnity payable (liability being otherwise admitted). However, the mixture of French and English insurers clearly agree to submit eventually to the jurisdiction of the French or English courts as applicable.

The leading insurers were jointly Commercial Union in London and Union des Assurances de Paris, with the policy having been placed jointly by Sedgwick Limited and Gras Savoye/Faugère & Jutheau.

Appendix 8

Sample Policy of Insurance for Employer's Loss of Liquidated Damges – JCT 80 Amendment 1986 clause 22D

In consideration of the Contractor and/or the Employer

(a) having made a proposal or supplied information which shall form the basis of this policy
(b) having paid or agreed to pay the Premium

to the Trinity Insurance Company Limited (hereinafter called the Company)

The Company agrees (subject to the terms, conditions, exceptions and limitations contained herein or endorsed hereon) that if during the Period of Insurance or any further period for which the Contractor or the Employer have requested cover and agreed to pay and the Company to accept the appropriate additional premium

Practical Completion of the Works be delayed directly in consequence of loss or damage by one or more of the Insured Perils as defined in the Specification forming part of this policy of or to

the permanent works, temporary works, unfixed materials and goods intended for incorporation in the Works (hereinafter referred to as the Property)

or any temporary buildings, plant, tools or equipment for use in connection with the Works

all whilst at or adjacent to the Site of the Works (loss or damage so caused being hereinafter termed Damage)

then provided that an extension of time therefor has been granted by the Architect in accordance with the Contract in consequence of the Damage the Company will pay to the Employer the Nominated Amount during the Payment Period less the Policy Excess in accordance with the provisions and definitions contained in the Schedule and Specification forming part of this policy

provided that

(1) at the time of the happening of the Damage there shall be in force an insurance on the Property arranged in accordance with the appropriate part of Clause 22A, B or C of the Contract and that liability for Damage thereto (where applicable) has been admitted thereunder (or would have been but for the operation of a proviso excluding liability for losses below a specified amount)

(2) the liability of the Company shall in no case exceed the Nominated Amount set out in the said Schedule nor in total the Sum Insured hereby (or such other sum or

sums as may hereafter be substituted therefor by memorandum signed by or on behalf of the Company)

On behalf of the Company

_____

THE SCHEDULE

(Forming part of Policy No.   )

The Employer:
The Contractor:
The Architect:
The Works:
The Site of the Works:

Practical Completion of the Works: as defined in the Contract

The Insured Perils: as detailed in the Specification attached hereto

The Nominated Amount: £   per
or such other lesser amount which shall be substituted by reason of Clause 18.1.4 of the Contract

The Sum Insured: £

Payment Period: the period during which Practical Completion of the Works is delayed directly in consequence of the Damage
(1) beginning with the date upon which but for the Damage Practical Completion of the Works would have been achieved and
(2) ending not later than
(a) the maximum period thereafter for which an extension of time has been granted or
(b) the Maximum Payment Period thereafter whichever is the less

Maximum Payment Period:

The Policy Excess: £   each and every occurrence

The Contract: The JCT Standard Form of Building Contract 1980 Edition entered into by the Employer and the Contractor for the Works

The Period of Insurance: From   to
both days inclusive

The Premium: £

THE SPECIFICATION

(Forming part of Policy No.   )

The Insured Perils:

1. Fire
2. Lightning
3. Explosion
   For the purpose of this insurance pressure waves caused by aircraft or other aerial devices travelling at sonic or supersonic speeds shall not be deemed explosion
4. Aircraft and other aerial devices or articles dropped therefrom excluding damage occasioned by pressure waves caused by aircraft or other aerial devices travelling at sonic or supersonic speeds
5. Earthquake
6. Riot Civil Commotion Strikers Locked-Out Workers or Persons taking part in labour disturbances or Malicious Persons acting on behalf of or in connection with any political organisation excluding damage resulting from cessation of work
   provided that full details of such Damage shall be furnished to the Company within seven days of its happening
7. Storm or Tempest or Flood excluding damage by frost
8. Bursting or overflowing of water tanks apparatus or pipes

EXCEPTIONS

This policy does not cover payments resulting directly or indirectly from any of the following causes:

(i)   war invasion act of foreign enemy hostilities (whether war be declared or not) civil war rebellion revolution insurrection or military or usurped power

(ii) confiscation requisition acquisition or destruction by order of any government or other authority

(iii) loss or destruction of or damage to any property whatsoever or any loss or expense whatsoever resulting or arising therefrom or any consequential loss directly or indirectly caused by or contributed by or arising from:

   (a) ionising radiations or contamination by radioactivity from any nuclear fuel or from any nuclear waste from the combustion of nuclear fuel
   (b) the radioactive toxic explosive or other hazardous properties of any explosive nuclear assembly or nuclear component thereof

(iv) loss or destruction of or damage to any property in Northern Ireland or loss resulting therefrom caused by or happening through or in consequence of:
   (a) civil commotion
   (b) an unlawful wanton or malicious act committed maliciously by a person or persons acting on behalf of or in connection with any unlawful association.

For the purpose of this Exception:

'Unlawful association' means any organisation which is engaged in terrorism and includes an organisation which at any relevant time is a proscribed organisation within the meaning of the Northern Ireland (Emergency Provisions) Act, 1973.

In any action, suit or other proceedings where the Company alleges that by reason of the provisions of this Exception any loss, destruction or damage is not covered by this policy the burden of proving that such loss, destruction or damage is covered shall be upon the Contractor and/or the Employer.

MEMORANDA

(The clause numbers mentioned are those specified in the Contract)

1. Payments made under this policy shall be subject to the Contractor in accordance with Clause 25.2 giving written notice to the Architect that the Works are likely to be delayed and the Architect granting an extension of time in accordance with Clause 25.3.

2. Where the Relevant Events stated by the Architect in accordance with Clause 25.3.1.3 are not limited specifically to the Damage insured by this policy the Company shall agree with the Employer that part of such extension which is due to the Damage and thereafter the Company shall pay to the Employer the Nominated Amount for the Payment Period so agreed.

3. Where agreement cannot be reached as to that part of an extension of time which has resulted from the Damage the Employer and the Company hereby agree to refer at the cost of the Company to an architect (other than the Architect named in the Schedule) whose decision shall be binding.

   The architect shall be a practising member of the Royal Institute of British Architects acceptable to both the Employer and the Company or otherwise as nominated by the president for the time being of that Institute.

4. Notwithstanding the Contractor's obligations under Clause 25.3.4.1 necessary and reasonable additional expenditure may be incurred with the approval of the Company for the sole purpose of avoiding or diminishing the payments which but for that expenditure would have been made under this policy but not exceeding the amount of the payments thereby avoided.

CONDITIONS

1. This policy shall be voidable in the event of misrepresentation misdescription or non-disclosure in any material particular

2. If at any time after the commencement of this insurance
   (a) the Contractor's or Employer's business be wound up or carried on by a liquidator or receiver or permanently discontinued
      or
   (b) the Contractor or Employer become bankrupt or make a composition or enter into any deed of arrangement with creditors
      or
   (c) the Contractor's or Employer's interest cease otherwise than by death
      or
   (d) any alteration be made either in the Works or the Property whereby the risk of Damage is increased

this policy shall be avoided unless its continuance be admitted by endorsement signed by or on behalf of the Company

3. The Contractor and Employer shall take all reasonable precautions to prevent loss or damage and the Company's representatives shall have access to the Site of the Works at all reasonable times

4. On the happening of any event giving rise or likely to give rise to a claim under this policy the Employer or the Contractor on the Employer's instructions shall

    (a) forthwith give notice therefore in writing to the Company

    (b) allow immediate access to the Site of the Works for the purpose of inspecting the Works and provide all evidence as may be required to a loss adjuster appointed by the Company

    (c) so far as may reasonably be practicable take precautions to preserve any things which might prove necessary or useful by way of evidence in connection with any claim

    (d) do and concur in doing and permit to be done all things which may be reasonably practicable to minimise the delay in the completion of the contract or to avoid or diminish loss

    (e) in the event of a claim being made under this policy at the Employer's expense deliver in writing to the Company not later than thirty days after the expiry of the Payment Period or within such further time as the Company may in writing allow a statement setting forth particulars of the Employer's claim together with details of all other insurances covering the Damage or consequential loss of any kind resulting therefrom

    (f) forthwith at the request and expense of the Company do and concur in doing all such acts and things as the Company may reasonably require for the purpose of enforcing any rights and remedies or obtaining relief or indemnity from other parties (other than the Contractor or any sub-contractor thereof against whom or which any existing rights of subrogation are waived) to which the Company shall be or would become subrogated upon its making payment under this policy whether such acts and things shall be or become necessary or required before or after payment by the Company.

    No claim under this policy shall be payable unless the terms of this condition have been complied with and in the event of noncompliance therewith in any respect any payment on account of the claim already made shall be repaid to the Company forthwith.

5. If any claim be in any respect fraudulent or if any fraudulent means or devices be used by the Employer or anyone acting on his behalf to obtain any benefit under this policy or if any Damage be occasioned by the wilful act of or with the connivance of the Employer all benefit under this policy shall be forfeited

6. If at the time of any Damage resulting in a loss covered by this policy there be any other insurance effected by or on behalf of the Employer covering such loss or any part of it the liability of the Company hereunder shall be limited to its rateable proportion of such loss.

7. If any difference shall arise as to the amount to be paid under this policy (liability being otherwise admitted) such difference shall be referred to an arbitrator to be appointed by the parties in accordance with the statutory provisions in that behalf for the time being in force. Where any difference is by this condition to be referred to arbitration the making of an award shall be a condition precedent to any right of action against the Company

8. Without prejudice to the Contractor's obligations pursuant to Conditions 3 and 4 of this policy the Contractor in taking out this insurance is acting as the agent of the Employer.

Appendix 9

Specification forming part of an advance profits policy

SPECIFICATION referred to in Business Interruption Policy No
in the name of

Item No	Sum Insured
1 On Anticipated Rent	£ ....

The Insurance under Item No 1 is limited to the loss sustained by the Insured in consequence of the Accident and the amount payable as indemnity thereunder shall be

(a) the loss of Anticipated Rent suffered by the Insured during the Indemnity Period
(b) the additional expenditure necessarily and reasonably incurred for the sole purpose of avoiding or diminishing the loss of Anticipated Rent which but for that expenditure would have taken place during the Indemnity Period but not exceeding the amount of the reduction in Anticipated Rent thereby avoided

less any sum saved during the Indemnity Period in respect of such charges of the Business that would have been payable out of anticipated Rent as may cease or be reduced in consequence of the Accident

provided that if the Sum Insured by this item be less than £....... the Annual Anticipated Rent the amount payable shall be proportionately reduced

Note 1 – Unless the Insured can provide evidence that it was the intention to lease or rent the Premises to one tenant only then should the Contract be completed and the Premises vacant at the time of the Accident the Company shall be liable for loss in respect only of the portion or portions of the Premises physically affected by the Accident

but in as far as the loss of Anticipated Rent in respect of

(i) the portion or portions of the Premises unaffected by the Accident
(ii) an Accident to portion or portions of the Premises not capable of direct occupation but providing services or access within the Premises

payment will be made on the provision of evidence that occupation of the Premises in whole or in part would have taken place during the Indemnity Period

Note 2 – If at the time of the Accident it is established that it was the Insured's sole intention to sell the leasehold or freehold of the Premises this Insurance shall remain valid and the Indemnity will be based on the application of the Percentage Rate of Interest per annum to the proposed selling price of the leasehold or freehold of the Premises (agreed or assessed by the Professional Valuer as appropriate) calculated

311

pro rata for the length of the Indemnity Period but not exceeding in all the indemnity that would have been granted if it had been the Insured's intention to lease or rent the Premises or the Sum Insured whichever shall be the less

In assessing the actual amount on which interest will be paid consideration will be given to any amounts saved by the Insured in consequence of the Accident

DEFINITIONS

The Business – Property Owners and Developers

The Premises – The site of the Contract and/or the completed buildings at

. . . .

The Contract – The erection and/or reconstruction and/or redevelopment of Buildings at the premises

Anticipated Rent (if evidence is provided of an agreement with a prospective tenant) – The money that would have been paid or payable to the Insured in respect of accommodation provided in course of the Business at the Premises

Anticipated Rent (if evidence is not provided of an agreement with a prospective tenant and the Contract is not complete) – The money that would have been paid or payable to the Insured in respect of accommodation provided in course of the Business at the Premises assessed by the Professional Valuer at the rates deemed to apply to the Premises at the date upon which but for the Accident the Contract would have been completed

Anticipated Rent (if evidence is not provided of an agreement with a prospective tenant and the Contract is complete) – The money that would have been paid or payable to the Insured in respect of accommodation provided in course of the Business at the Premises assessed by the Professional Valuer on a charge for area basis at the rates deemed to apply to the Premises at the date of the Accident

Indemnity Period (if evidence is provided of an agreement with a prospective tenant) – The period beginning with the date upon which but for the Accident the Premises would have been let and occupied and ending not later than   months thereafter during which the results of the Business shall be affected in consequence of the Accident

Indemnity Period (if evidence is not provided of an agreement with a prospective tenant and the Contract is not complete) – The period beginning with the date upon which but for the Accident the Contract would have been completed and ending not later than   months thereafter during which the completion of the Contract is delayed in consequence of the Accident

Indemnity Period (if evidence is not provided of an agreement with a prospective tenant and the Contract is complete) – The period beginning with the occurrence of the Accident and ending when that part of the Premises capable of direct occupation and affected by the Accident is restored to its predamaged condition but not later than   months after the occurrence of the Accident

Indemnity Period (if the terms of Note 2 apply) – the three previous definitions shall apply as appropriate except that the word "purchaser" is substituted for "tenant" and the word "sold" is substituted for "let and occupied"

Annual Anticipated Rent – The proportional equivalent for a period of 12 months of the Anticipated Rent (as defined) during the Indemnity Period

Percentage Rate of Interest – The actual rate of Interest payable by the Insured during the Indemnity Period in respect of Capital borrowed to finance the Contract adjusted in respect of the non-borrowed portion of the proposed selling price of the leasehold or freehold of the Premises to the 90-day money market rate pertaining during the Indemnity Period

The Professional Valuer – A practising member of the Royal Institution of Chartered Surveyors whose appointment shall be satisfactory to both the Insured and the Insurers or otherwise by nomination of the President for the time being of the Royal Institute of Chartered Surveyors

Note – The fees payable to the Professional Valuer shall be paid by the Company

Memo: 1 – In the settlement of any loss under this Policy account will be taken of any factor which might affect the trend of the business and consideration will be given to any variations in or special circumstances affecting the business either before or after the Accident or which would have affected the business had the Accident not occurred so that the final settlement of the loss shall represent as nearly as may be reasonably practicable the results which but for the Accident would have been obtained during the Indemnity Period in accordance with the terms of this Policy

Memo: 2 – If during the Indemnity Period in consequence of the Accident alternate premises be made available to secure to the Insured Rental or other income which would otherwise have been received by the Business at the Premises account shall be taken thereof in arriving at the Indemnity hereunder

Memo: 3 – Payments on account will be made to the Insured monthly during the Indemnity Period if desired

Memo: 4 – Any particulars or details contained in the Insured's books of account or other business books or documents which may be required by the Insurers under Condition 3 of this Policy for the purpose of investigating of verifying any claim hereunder may be produced and certified by the Insured's Auditors and their certificate shall be prima facie evidence of the particulars and details to which such certificate relates

The Company will pay to the Insured the reasonable charges payable by the Insured to their Auditors for producing such particulars or details or any other proofs information or evidence as may be required by the Company under the terms of Condition 3 of this policy and reporting that such particulars or details are in accordance with the Insured's books of account or other business books or documents

provided that the sum of the amount payable under this clause and the amount otherwise payable under the policy shall in no case exceed the total sum insured by the policy.

Appendix 10

ABI suggested policy exclusions for clause 21.2.1 insurance

'Model' Exclusions

The indemnity will not apply in respect of

1. injury or damage to property
    a) caused by the negligence omission or default of the Contractor his servants or agents or of any sub-contractor his servants or agents
    b) attributable to errors or omissions in the designing of the Works
    c) which can reasonably be foreseen to be inevitable having regard to the nature of the work to be executed or the manner of its execution
    d) which is the responsibility of the Employer under the provisions of Clause 22C.1 of the JCT Standard Form of Building Contract (1980 Edition) or any equivalent thereof

2. damage to the Works and Site Materials brought on to the site of the Contract for the purpose of its execution except in so far as any part or parts thereof are the subject of a practical completion certificate

3. any costs or expenses incurred by the Employer in respect of liquidated damages or any other sum payable by way of damages for breach of contract except to the extent that such costs or expenses would have attached in the absence of any contract

4. damage to property directly occasioned by pressure waves caused by aircraft or other aerial devices travelling at sonic or supersonic speeds

5. any expense liability loss claim or proceedings of whatsoever nature directly or indirectly caused by or contributed to by or arising from
    a) ionising radiations or contamination by radioactivity from any nuclear fuel or from any nuclear waste from the combustion of nuclear fuel
    b) the radioactive toxic explosive or other hazardous properties of any explosive nuclear assembly or nuclear component thereof

6. any consequence of war invasion act of foreign enemy hostilities (whether war be declared or not) civil war rebellion or revolution insurrection or military or usurped power

7. damage directly or indirectly caused by or arising out of pollution or contamination of buildings or other structures or of water or land or the atmosphere happening during the period of insurance

314

this exclusion shall not apply in respect of pollution or contamination caused by a sudden identifiable unintended and unexpected incident which takes place in its entirety at a specific moment in time and place during the period of insurance provided that all pollution or contamination which arises out of one incident shall be considered for the purposes of this insurance to have occurred at the time such incident takes place.

Note: a) Exclusion 2 may be incorporated as part of an Operative Clause

Period of Insurance

All 21.2.1 policies should include a Period of Insurance. This should represent the Contract Period plus a normal Maintenance or Defects Liability period (usually 12 months).

Appendix 11

The principal/contractor relationship at common law

The common law position as regards the relationship of principal and contractor is set out here, but the position between main contractor and sub-contractor is the same, (see *Maxwell* v. *British Thomson-Houston Co Ltd* (1902)). Usually these exceptions apply when a *personal* duty rests on the main contractor and it is no defence that the duty was delegated to the sub-contractor.

(1) Where the contractor is employed to do illegal work. In *Ellis* v. *Sheffield Gas Consumers Co* (1853) the defendants, without obtaining the necessary special powers, employed a contractor to open trenches in the streets of Sheffield. The plaintiff sustained injuries by falling over a heap of stones which had been left by the contractors, and the gas company was held liable for the contractor's negligence.

(2) Where the contractor is employed to do work which involves strict liability:

- Cases involving the rule in *Rylands* v. *Fletcher*. This case itself concerned liability for an independent contractor. Liability for fire is similar to liability under the rule in *Rylands* v. *Fletcher*. In fact the liability extends to cases where the fire is deliberately lit and negligently allowed to get out of control by an independent contractor (see *Balfour* v. *Barty King* (1957) and *H & N Emanuel Ltd* v. *Greater London Council* (1971)).

- Withdrawal of support to land or buildings. In *Bower* v. *Peate* (1876) the defendant employed a contractor to pull down and rebuild his house. The contractor expressly undertook to support the plaintiff's adjoining house but this was damaged. The defendant was held liable for the damage done. The decision in this case could have been justified on the narrow grounds of nuisance or interference with an easement (see the law on this aspect in Chapter 9) and it has been included under the above heading for this reason. However, the court in fact based their decision on the wider principle that there is 'good ground for holding [the party authorising the work] liable for injury caused by an act certain to be attended with injurious consequences, if such consequences are not in fact prevented, no matter through whose default the omission to take the necessary measures for such prevention may arise'. This principle has, however, been criticised.

- Operations on or adjoining a highway creating dangers thereon. In *Tarry* v. *Ashton* (1876) the defendant employed a contractor to repair a lamp fixed to his house and overhanging the highway. It was not securely fixed and fell on the plaintiff, a passer-by. The defendant was liable because it was the defendant's duty to make the lamp reasonably safe. The contractor's failure to do this meant that the defendant had not performed his duty, and had created a public nuisance.

(3) Where the principal retains a measure of control over the contractor by the provision of machinery or men or both. In *Levering and Doe* v. *St Katharine's Dock Co* (1887) sub-contractors undertook to discharge ships with their own gangs, the members of which were on the books of the dock company as permanent workmen. The company provided the gear, and when two members of the gangs were injured, the company were held liable as having retained a sufficient degree of control over the gangs, although insufficient to make them liable as employers. With the tendency to use 'labour-only' sub-contractors, this situation may arise more often between main and sub-contractors.

(4) Where the principal is under a statutory duty to perform work in a particular manner. In *Hole* v. *Sittingbourne Railway* (1861) the defendants were authorised by statute to construct a bridge across a navigable river with a provision that the bridge should not detain any vessel navigating the river longer than was necessary. Faulty construction by independent contractors prevented the opening of the bridge for several days and the defendants were held liable.

(5) Where liability attaches due to the principal's own negligence. An example would be delegating the work to be carried out to an inexperienced or incompetent contractor or giving insufficient instructions to him. In *Robinson* v. *Beaconsfield Rural District Council* (1911) the respondent council were under a statutory duty to cleanse cesspools. It delegated the cleansing of certain pools to a contractor but failed to give to him any directions as to the disposal of the filth. The contractor deposited the filth on the appellant's land. It was held that the respondents were liable for their failure to take proper precautions to dispose of the sewage.

(6) Unusually hazardous work. In *Honeywill & Stein Ltd* v. *Larkin Bros Ltd* (1934) the plaintiffs employed the defendants as independent contractors to take photographs by flash light. The defendants set fire to a theatre and the plaintiffs paid for the damage and attempted to recover from the defendants. The Court of Appeal held that the plaintiffs were liable as they had arranged for work to be done which involved some degree of special danger. It seems to be very much a matter of opinion as to what is work which in its very nature involves a special danger to another. In the comparatively recent case of *Alcock* v. *Wraith* (1993) the owner of one of a row of terraced houses with a continuous tiled roof, engaged a contractor to carry out re-roofing work. The contractor inadequately constructed the join between the new tiles in his employer's roof and the existing tiles on the neighbour's roof, with the result that the neighbour's roof leaked. The court held that an employer is liable for the tort of an independent contractor where there is a special risk or where the work from its very nature is likely to cause danger or damage.

Appendix 12

'Part A Introduction' from the JCT *Guide to Terrorism Cover*

4 As more fully described in Part B, insurers in 1992 were informed that their reinsurers would not be prepared to reinsure in respect of fire and explosion damage caused by terrorism. Insurers therefore informed the Government that without such reinsurance full cover for commercial and industrial buildings and for building works thereto for damage by fire or explosion caused by terrorism would be withdrawn. At the end of 1992 the Government agreed to act as a "reinsurer of last resort" and a new method of providing for such damage by fire or explosion caused by terrorism ("terrorism cover") was established: in summary "terrorism cover" would be excluded from policies but could be bought back by payment of standard premiums fixed for all policies and graded by reference to zones in the UK, and such premiums would be paid into a pool administered by a reinsurer established by the Government, Pool Reinsurance Company Limited ("Pool Re."). The Tribunal noted these arrangements and issued advice to its constituent bodies as follows:

(a) As JCT Contracts require either the Employer or the Contractor to insure for Specified Perils (e.g. JCT 80, clause 22C.1) or for All Risks (e.g. JCT 80, clause 22A or 22B or 22C.2) which would include insurance for loss or damage due to fire or explosion however caused (other than by the defined "excepted risks" which do not include terrorism) there was no need to refer specifically to cover for fire or explosion damage caused by terrorism ("terrorism cover");

(b) While the new arrangements mean that the Employer or the Contractor, in order to meet their insurance obligations under the Contract, would have to buy back "terrorism cover" it was not considered desirable to amend the JCT Forms to provide for such "buy back" to be optional;

(c) Constituent bodies are reminded that if the Employer or the Contractor, as the case may be, did not buy back "terrorism cover" they would be in breach of their contractual obligation to insure for damage due to fire or explosion however caused (other than by the defined "excepted risks").

5 The arrangements with Pool Re. are covered by an agreement (laid before Parliament on 30 July 1993) between the Government and Pool Re. pursuant to the Reinsurance (Acts of Terrorism) Act 1993. The Explanatory Summary of the Act's provisions issued by the DTI states:

"The Agreement (Government and Pool Re.) is of indefinite duration but may be terminated with effect from midnight 31 December in any year by either party giving at least 120 days notice in writing. The Secretary of State is also entitled to terminate the Agreement immediately in the event of Pool Re. failing to comply with the terms and conditions of the Agreement or the Company's Memorandum and Articles of Association being amended without prior consent. The Agreement may similarly be terminated immediately if the performance of the Agreement becomes impossible or if Pool Re. is wound up. In these latter cases only, the Government will continue to honour claims arising from Reinsurance Agreements in force at the date of termination."

While a Government Minister has given an assurance of the Government's objective to ensure the continued availability of terrorism insurance and to continue to stand behind the arrangements pursuant to the 1993 Act until there is adequate alternative cover, the Tribunal has had to take note of the Government's right of termination in its Agreement with Pool Re. As a consequence of such right of termination no insurer is prepared to provide terrorism cover without a limitation on the period of its availability. The Tribunal thus noted that neither the Employer nor the Contractor could rely on the "terrorism cover" being renewed. The position on a contract could therefore be that, before the date of issue of the Certificate of Practical Completion of the Works and without any default by the parties, there might be no terrorism cover for the Works and, where the Works are in or an extension to an existing structure, for the existing structure and its contents.

6   The Tribunal therefore confirmed its earlier advice set out in paragraph A4 that, in order to comply with the conditions set out in their various Forms, terrorism cover must be obtained; but has prepared Amendments for each of the Forms to deal with the situation where, before the date of issue of the Certificate of Practical Completion of the Works, "terrorism cover" is no longer available.

7   Part B describes in more detail the arrangements effective from 1 July 1993 on obtaining terrorism cover. Part C describes the effect of the Amendments referred to in paragraph A6 above. Appendix A set out the full text of the Amendment TC/94 to the Standard Form of Building Contract 1980 Edition (JCT 80). The 'TC' Amendments are to be included in each copy of the relevant JCT Form sold.'

Table of cases

Note

The following abbreviations of reports are used:

AC	Appeal Cases
All ER	All England Law Reports
App Cas	Law Reports (New Series) Appeal Cases
BLR	Building Law Reports
Bos & P	Bosanquet and Puller's reports
CILL	Construction Industry Law Letter
Ch	Chancery Division
Ch D	Law Reports (New Series) Chancery Division
CL	Current Law
Com Cas	Commercial Cases
Con LR	Construction Law Reports
DLR	Dominion Law Reports
Ex	Exchequer Division
E & B	Ellis and Blackburn
F & F	Foster and Finlason
H & N	Hurlstone and Norman
HL Cas	House of Lords Cases
IR	Irish Reports
ILR	Insurance Law Reports
KB	Law Reports, King's Bench
Mod	Modern Reports
Lloyds Rep	Lloyds Law Reports
Ll LR	Lloyds List Law Reports
LR HL	English and Irish Appeals
LT	Law Times Reports (New Series)
NZLR	New Zealand Law Reports
NPC	New Property Cases
NY	New York
QB	Law Reports, Queen's Bench
QBD	Law Reports, Queen's Bench Division
RTR	Road Traffic Reports
SJ	Solicitors' Journal
TLR	Times Law Reports
VLR	Victoria Law Reports, Australia
WLR	Weekly Law Reports

Index

(**Note:** App = Appendix)